Felicity Lewis (left) is the national explainer editor for *The Age* and *The Sydney Morning Herald*. She has worked on titles ranging from *The Independent* in London to *The Age (Melbourne) Magazine*. As national multimedia editor, her team's awards included a Walkley for production in 2017. She edited the anthologies *Explain That* (2021) and *What's It Like to Be Chased by a Cassowary* (2020).

Jackson Graham (centre) has been a reporter on the Explainer desk at *The Sydney Morning Herald* and *The Age* since 2023. He started in journalism on regional newspapers in Victoria. Several of his Explainers, including the one on autopsies in this book, were highly commended for science writing at the Melbourne Press Club's Quill Awards in 2024.

Angus Holland (right) has been an explainer reporter for *The Age* and *The Sydney Morning Herald* since 2023. His previous roles at *The Age* include managing editor, opinion editor and editor of *The Age (Melbourne) Magazine*. He is co-author of the bestselling title *100 Great Businesses and the Minds Behind Them*.

WHY DO PEOPLE QUEUE FOR BRUNCH?

The EXPLAINER guide
to modern mysteries

Edited by Felicity Lewis

ALLEN&UNWIN

SYDNEY · MELBOURNE · AUCKLAND · LONDON

Allen & Unwin
Cammeraygal Country
83 Alexander Street
Crows Nest NSW 2065
Australia
Phone: (61 2) 8425 0100
Email: info@allenandunwin.com
Web: www.allenandunwin.com

Allen & Unwin acknowledges the Traditional Owners of the Country on which we
live and work. We pay our respects to all Aboriginal and Torres Strait Islander
Elders, past and present.

A catalogue record for this
book is available from the
National Library of Australia

ISBN 978 1 76147 182 7

Chapter opener illustrations by Josh Durham, Design by Committee
Set in12/16 pt Sabon LT Std by Midland Typesetters, Australia
Printed and bound in Australia by the Opus Group

10 9 8 7 6 5 4 3 2 1

The paper in this book is FSC® certified.
FSC® promotes environmentally responsible,
socially beneficial and economically viable
management of the world's forests.

Contents

Introduction

My mother, Dianne, was a frank and practical person, even when she was dying of cancer. 'You're looking well,' a boutique owner told her one day. 'Thank you,' Mum chirped, then added, 'Actually, I've just been told I've got a few short months to live.' Even in her final days, as we lay chatting on her bed, she said, 'Quick, hand me my iPad, I just realised I need to give you girls my frequent-flyer points—I can't take them with me!'

It's 10 years now since Mum died. Like many people, I try to keep up with reports about new and improved cancer treatments; and my heart sinks when I learn that, amidst all of the advances, yet another person—a relative, a friend, a colleague—has had their life up-ended by a cancer diagnosis. Why is it so hard to cure?

As it happens, it's my job to find answers to such questions. I'm the national explainer editor at *The Age* and *The Sydney Morning Herald*, commissioning in-depth articles that demystify some of life's curlier matters.

Explainers, a media format dealing with the how and why of events, are not new. But the fast and fragmented ways we catch up on news in a digital age increasingly underline their value: not only as fascinating reads but as companion pieces, offering context and background to events as they unfold.

Beyond the news cycle, our Explainers also unpack issues that can confront us in our own lives. We tackle subjects we find intriguing, interrogating them from many angles, and through granular research. 'It has so many human elements,' a reader said of our Explainer on how to build sandcastles, 'philosophy, engineering, chemistry, physics'. In short, when I first meet someone and they ask me what I do, I say, 'I'm part of a team of journalists who get to explore whatever makes us curious'.

Ideas for Explainers can surface slowly, as with the one about cancer, or without warning. Over coffee one day, a neighbour who once sailed on cargo ships told me about rogue waves. 'You should explain those,' he said. At a barbecue, friends' speculation about what they'd do if they ever retired—imagine that!—led to an Explainer article on this phase of life. At some point, 'narcissist' became a trendy put-down, but what does it really mean? And long queues of shiny, happy people waiting for brunch outside certain cafes near my home prompted me—who hasn't queued since I bought tickets to a Spandau Ballet concert in 1985—to wonder, why?

Of course, following lines of inquiry tends to lead to still more questions. It was while she was working on an Explainer about the Moon that reporter Sherryn Groch learned of two mysterious blobs discovered in the Earth's core. This inspired her to take us deep below our planet's crust—read on to find out what lies beneath. Explainer journalist Jackson Graham had scrawled 'search and rescue' and 'grave fears' in his notepads many times as a country news reporter; in this book, he investigates how search and rescues unfold, through a nail-biting account of a woman lost in the wilderness. And, as another festive season loomed, our other Explainer team member, Angus Holland, knew he'd soon be hobnobbing with people he hadn't seen in ages. 'I wondered,' he told me, 'How can you have a conversation with virtually anybody that's more enjoyable?' We like this top tip from Ancient Rome: avoid ho-hum topics such as gladiators!

Journalists who work in specialist fields in our newsrooms also bring their expertise to these chapters. Transport reporter Patrick Hatch examines what makes car crashes so deadly (it's not always just bad driving). Senior writer Matt Wade shows why India becoming the world's most populous nation—it overtook China in 2023—is but a blip amid huge demographic shifts reshaping our world. Style editor Damien Woolnough presents a brief history of a surprisingly controversial tog.

No AI chatbots were used in the making of this anthology. Any phenomena that might seem weird—tiny magnetic crystals, copper-encrusted sea vents, dancing bees—*are* weird but they're not hallucinations, they're real. Our information was gathered by chatting with humans—lots of them, including experts around the world, from Liverpool to Washington to Utrecht, and everyday people at the heart of our stories. Our Explainer on retirement, for example, contains wise counsel from generous readers who put up their hands to be interviewed by us. Our Explainer on the vital (and secretive) microchip industry includes rare insights from workers in Taiwan's 'Silicon Valley' who agreed to speak with our correspondent Eryk Bagshaw.

You'll notice an adventure theme, too. We call it the Jules Verne effect. Apart from taking you on a journey to Earth's centre, we fly you to the Moon and plunge you many leagues under the sea. For those of you who prefer to chill on the beach: did you know the world's tiniest sandcastles are etched on grains of sand? Or that budgie smugglers got someone arrested at Bondi in 1961? For more captivating facts, read on—and feel free to recycle them in social situations. After all, as one of our experts tells us, small talk might seem trivial but it is actually a way to connect with another person before you discuss something deeper or—dare we say it?—controversial. Enjoy.

Felicity Lewis

1

WHAT'S A ROGUE WAVE?

Scientists thought these monster waves were a myth. Now they're racing to understand them.

Sherryn Groch

It came out of the storm: a sudden wall of water as tall as a 10-storey building. On deck, explorer Jules Dumont d'Urville estimated the wave loomed at least 30 metres high—and it was bearing down fast on his ship, the *Astrolabe*. Somehow, they made it back to shore, losing just one man on that dangerous crossing of the Indian Ocean in 1826.

But when Dumont d'Urville, known as France's Captain Cook, and his crew later recounted the tale of the monster wave, no one believed them. As far as the scientists of the 19th century were concerned, what the sailors had seen was impossible: no wave could reach more than nine metres in height. For centuries, ships' disappearances at sea were blamed on pirates or misadventure, and stories of giant waves were dismissed as readily as legends of sea monsters.

Then in 1995, a sensor on a Norwegian oil rig captured proof of what Dumont d'Urville had faced: a wave 26 metres tall, more than twice the size of any recorded in the area in the hours before, taller even than the hypothetical waves that scientists then believed could happen only once every 10,000 years. That same year, when the ocean liner *Queen Elizabeth 2* (*QE2*) was struck by a 27-metre-high wave in the North Atlantic, scientists had to admit something else: these so-called rogue waves weren't just possible, they happened relatively frequently. Facing down that 1995 wave, the *QE2* captain said it looked as if they were headed for the white cliffs of Dover.

So, what are rogue waves exactly? And are we getting any better at predicting (and surviving) them?

ARE ROGUE WAVES TSUNAMIS?

Today, the monster wave of sailors' nightmares has a formal scientific definition: a rogue wave is at least twice as high as recent waves around it. It can rise and disappear quickly out of a stormy sea but it can come out of nowhere too—in calm waters. Professor Nail Akhmediev, a theoretical physicist at the Australian National University, has been trying to use equations to predict rogue waves. He says survivors will sometimes describe otherwise calm weather before the monster wave appears. These waves can even swallow rescue helicopters that swing down to the water to help seafarers in trouble, as one coast guard chopper discovered off the coast of Alaska in 2004. Witnesses described how, as crew members were being airlifted from a stricken freighter, a wave 'larger than any encountered before' struck the bow and engulfed the chopper as well. (All the chopper crew were eventually saved but six of the ship's crew who had been on the chopper died.)

In more than 40 years at sea, marine engineer Karsten Petersen says he never saw a more terrifying wave than the one that crashed into the chemical tanker *Stolt Surf* during a voyage across the Pacific Ocean from Singapore to the United States in 1977. From the bridge, he managed to photograph it—snapping some of the few images ever taken of a rogue wave.

Petersen, now retired, recalls the heart-stopping moment when the water crashed on deck, more than 22 metres above sea level—and they weren't sure if the ship was underwater or still afloat. Looking out of the bridge windows was 'like looking into a water tank,' he says.

The water crashed on deck, more than 22 metres above sea level.

'No sky, no horizon, no ship in front of you—only water.' But 'like a miracle', the white mist cleared, and the tanker pressed on through the wild seas, although with heavy damage.

Peter van Duyn also spent many years sailing the world: captaining ships, weathering hurricanes and cyclones and skirting icebergs, in the 1970s and '80s—before rogue waves were accepted science. 'I've seen some big, big waves,' he says. 'But back then there was no proof and it is hard to judge the exact size on the bridge of a ship. When you got back on land, they'd just say you were dreaming.'

Rogue waves are not tsunamis, which are triggered by a large displacement of water following an event such as an earthquake, volcanic eruption or landslide. They affect the entire water column. At sea, you might not even notice a tsunami wave rolling under you but near the shore, as it enters the shallows, it can climb to terrifying heights, often kilometres wide.

Rogue waves, on the other hand, are generally thought of as moving across the surface, although Akhmediev says they can also form deep below, sometimes called rogue internal waves. It's believed an internal wave in the Bali Sea tore apart an Indonesian submarine and killed all 53 on board in 2021. It may have been up to 100 metres tall. 'There was no other explanation,' says Akhmediev of the tragedy. 'They found the sub split into three parts on the bottom of the sea but it hadn't been attacked by anything or anyone.' The area is a known hotspot for such sea turbulence, scientists say, and satellite images taken at the time revealed waves on the surface, likely 'ripples' from a giant wall of water surging below.

Aside from eagle-eyed sailors, Akhmediev says, there are now many ways to detect rogue waves, from measuring pressure at the bottom of the ocean to special buoys that gauge wave heights. He estimates there are at least 10 of them at any one time in the ocean. 'Of course, luckily, there's not that many ships out there [compared to the vast ocean], so not many will encounter them.'

In 2004, scientists using satellite data from the European Space Agency spotted at least 10 significant rogue waves, each 25 metres or higher, within just three weeks. At the time, the agency said rogue waves probably sank most of the 200 supertankers and container ships over 200 metres long that had gone down in severe weather over the previous two decades.

Sometimes, in a phenomenon known as the 'three sisters', giant waves will strike in threes. In 2010, two people were killed when three rogues hit the *Louis Majesty* cruise ship off the coast of Spain. 'The first didn't do much damage but the second and third blew out the glass and flooded into multiple decks,' says Akhmediev. Multiple rogue waves were to blame for the 1998 Sydney to Hobart yacht race tragedy in which six lives and five boats were lost in wild conditions.

Then in 2012, Akhmediev and his colleagues proved the existence of another oddity: rogue holes, the inverse of a rogue wave, where the depth of the trough (the wave's lowest point) can be twice the size of the height of its crest. 'So they can be even steeper than the rogue wave, and very dangerous too—this great hole opening suddenly in the sea.'

HOW DO ROGUE WAVES FORM?

If forecasting the weather is complicated, oceans are an even more complex beast, Akhmediev tells us. The wind whips up waves, driving them across the seas for thousands of kilometres. But everything from the geography of coastlines and the ocean floor to the movements of the Earth and Moon— even the amount of salt in the swells—can affect how these waves form.

Still, there are two main schools of thought to explain the physics of rogue waves. The linear theory argues that when two wave crests meet they can merge into a single, bigger wave (just as a trough meeting a crest can cancel it out and flatten the sea). Sometimes different columns of waves 'or wave trains' will collide, often when different currents run into each other, forming huge waves for short periods. 'Think of cars travelling at speed,' says Akhmediev. 'Every now and then, there's a pile-up.'

At certain hotspots, scientists can see this in action. The most infamous is off the southeast coast of Africa, says van Duyn, where the fast-moving Agulhas current collides with waters from the Indian and Southern oceans. These forces can have an amplifying effect on the waves, making them steeper, like focusing light from a magnifying glass.

When scientists at Oxford recreated in a tank the wave that hit the oil rig near Norway in 1995, known as the Draupner wave, they saw something that looked remarkably similar to the great wave depicted by Japanese artist Katsushika Hokusai in his iconic 19th-century print *Under the Wave Off Kanagawa*. But they also found that the dynamics of how waves break change when two peaks cross. If they meet at the right angle, sometimes they can combine to form a rogue.

The linear theory does not explain many elements of waves, though, including why some rogues form on seemingly calm waters. 'And,' says Akhmediev, 'we know waves don't act in a nice linear way,' where the size of a rogue is in perfect proportion to the waves that came together to form it. That's why he and many mathematicians are looking to the strange world of quantum physics for an explanation. 'All particles act like waves, after all, even on a subatomic level,' he says.

Rogue waves can be more than five times the size of others around them.

Under their non-linear theory, waves not directly interacting can sometimes share energy. 'Like cars, they carry enormous amounts of energy,' Akhmediev notes. 'And sometimes it can grow', leeching out from other surrounding waves and 'concentrating into a single rogue'.

In 2012, he and German scientists tested this theory in the lab by generating a freak wave in otherwise calm water that capsized a Lego pirate ship. ('Our pirate survived it, though. He now lives on my desk,' Akhmediev laughs.) The experiment revealed that rogue waves can be even bigger than previously thought, more than five times the size of others around them. The scientists dubbed them 'super rogue waves'.

HOW DO YOU SURVIVE A ROGUE WAVE?

Sailors suddenly confronted with a rogue have few good choices, says van Duyn, who is now a maritime expert. If a giant wave smashes on deck, it can conk out a ship's engines and other systems, or wash away its crew and cargo. Even worse, he says, is a rogue wave at night. 'If you can't see it coming, you don't have any chance to steer the ship.'

Ideally, van Duyn says, you should sail head-on into such a wave. Being hit from the side risks capsizing. Of course, going bow-first up that steep cliff of water comes with its own risks. If the wave is big enough, it could tear the ship apart. That's why ships wrecked by rogue waves are often found in pieces; in some cases, the force of the water is strong enough to punch holes through steel hulls.

'Sometimes when ships disappear, they're found completely broken in two,' says van Duyn. 'It doesn't happen as

Ships wrecked by rogue waves are often found in pieces.

often now, as we build ships better. Even some of the wild waves I've been up against, I've never really thought we were about to sink.' When van Duyn was sailing in the 1980s, at least one ship vanished every day. Today, the global fleet is far larger but losses have declined as design improves.

But experts, including Akhmediev, warn that ships still aren't built to withstand the force of rogue waves. Many bulk carriers are designed to weather waves only about 11 metres high, yet a review by sailor and researcher Craig Smith found every ship was likely to encounter at least one 20-metre wave over a 25-year lifetime.

'You can design ships better but obviously that's going to cost,' says van Duyn. 'And you can't design for everything. They thought they'd made the *Titanic* unsinkable with all those extra compartments and bulkheads, but she still went down. Or maybe you'll end up making it too strong, or too heavy.' Steel on a ship has to be flexible, not just strong, to move with the sea. From the bridge, you often see it flexing at the bow. 'It would snap otherwise,' van Duyn says. 'Ships are designed to roll to a certain extent too in storms, but once they are bent over at 40, 50 degrees, they start to take on water. I've lost steering before on a ship, and we couldn't keep it head-on to the wave. Things got precarious but we managed to right it.'

Ageing ships, as well as improperly loaded cargo or inferior steel, can lower the odds of surviving a rogue wave, he says. So can decisions made at sea. Ships often opt to slow down and ride out bad weather but when they are ferrying goods on a tight schedule, time is money. When the Suez Canal was blocked for six days in 2021 by a 400-metre-long container vessel that ran aground, for example, ships instead braved the turbulent waters around southern Africa's Cape of Good Hope to meet their deadlines. Many shipping companies also decided to take the long way round when, in 2023, Houthi forces in Yemen began attacking ships in the Red Sea. 'They had to go into the rogue wave territory we usually avoid,' says van Duyn.

ARE WE GETTING BETTER AT PREDICTING ROGUE WAVES?

Even though scientists are still debating the (likely multiple) causes of rogue waves, they have started trying to predict them. 'We're really at the beginning of that now,' says Akhmediev. In the United States, the National Oceanic and Atmospheric Administration is working on an hourly forecast for potentially hazardous ocean conditions called WAVEWATCH III.

Other mathematicians argue for calculating and then charting the most efficient way rogues can form, to factor in both the linear and non-linear theories. Trials of this approach in wave tanks have been fairly accurate, although lab conditions can never match the real-world chaos that sailors encounter.

Whatever the algorithm, the trick is making predictions fast enough to be of use to ships. On the high seas, conditions can shift minute to minute. Akhmediev points again to the example of street traffic. It might be fairly easy to measure a car's speed and the distance it has travelled, but to predict exactly where that car will be in an hour's time, factoring in traffic lights, other cars, weather and more, is 'very tricky maths indeed'. Science pulls it off (to some extent) with weather forecasts, but he says the sea is even slippier to divine than the atmospheric rivers on high. To predict rogue waves, you need to know 'in detail the initial conditions of the sea, so you have to be scanning all the nearby waves somehow', he says.

One solution posited by Akhmediev and his international collaborators is a device on each ship that would continually scan conditions and calculate risk factors. 'Or maybe we could watch from space and relay that down to many ships at once,' he says. 'Clouds do make that difficult, though.'

Meanwhile, as climate change fuels wilder storms across the globe, this will lead to bigger rogue waves in some places, scientists warn. In areas such as the Arctic, where melting ice is opening up new shipping routes, wave heights are tipped to climb by six metres. Australia too will likely see its waves grow by 15 per cent if the world warms by two degrees Celsius.

Data collected from buoys off the western seaboard of the United States suggest rogue waves may be happening less often as storm systems shift, but are reaching bigger heights when they do form. In the southern hemisphere, University of Melbourne oceanographer Professor Ian Young has been studying the impact of climate change on waves for more than 30 years and has found that, on average, waves in the volatile Southern Ocean have grown by about 30 centimetres since 1985 as extreme winds become more frequent. University of Melbourne engineer Professor Alessandro Toffoli has also set sail in those fierce seas, mapping the ocean in 2017 as it was whipped up by storms. In those conditions, Toffoli says, his team recorded one rogue wave every six hours, suggesting wind may play a bigger role in the phenomenon than previously thought.

The science of rogue waves has come a long way since van Duyn's sailing days. 'But a lot of it will still come down to good seafaring,' he says. 'And hoping, when you see it, you get the chance to grab the wheel.'

2

WHERE DID BUDGIE SMUGGLERS COME FROM?

Once they could get you arrested; now they're a swimwear staple. A brief history of a controversial tog.

Damien Woolnough

With Holden Commodore key rings largely lost to the past, symbols of Australian manhood don't come smaller than swimming trunks. Our love for briefer-is-better styles can be found in celebrated artworks such as Charles Meere's 1940 painting *Australian Beach Pattern*, in movies like *Coolangatta Gold* and poolside at the country's most memorable sporting triumphs—as well as beachside any day of the year.

The rise in popularity of labels such as Budgy Smugglers, aussieBum and Sluggers, along with the enduring success of Speedo, has kept brief styles in the faces of Australian beachgoers, and their presence remain a perennial talking point.

Where did these skimpy swimming costumes come from? How did they become so, um, big in Australia? And what does their future hold?

WHO INVENTED BUDGIE SMUGGLERS?

Budgerigars are small, long-tailed parakeets that have lived in the wilds of Australia for several million years. They're famous for their pretty plumage, for being fashionable as pets and for lending their name to a type of tog. The *Oxford English Dictionary* finally recognised the term 'budgie smugglers' in 2016, defining them as 'men's brief, tight-fitting swimming trunks'. The dictionary traces an early use of the term to the television mockumentary *The Games*, about the 2000 Sydney Olympics. In one scene, a character played by satirist and actor John Clarke discusses the no-fuss style of Des Renford, a Sydneysider who swam the English Channel 19 times. 'Des Renford would regularly take on the English Channel, Bryan,' he says. 'He would drop his tweeds, pull on a pair of oversized budgie smugglers, and he would drop a bomb off the white cliffs of Dover and start rolling his arm over.'

Of course, there are many variations on 'budgie smuggler': depending on whether you grew up swimming at Manly, Burleigh, Cottesloe or Gunnamatta, you might also call them sluggos, dick stickers, DPs, cluster busters, banana hammocks or lollybags. The point is, most people will understand that they need to brace for an eyeful when you tell them you're donning smugglers.

There are many variations on 'budgie smuggler', depending on whether you grew up at Manly, Burleigh, Cottesloe or Gunnamatta.

In 1927, a charismatic Swede named Arne Borg—a rival of Sydney swim star Andrew 'Boy' Charlton—put on a dazzling display at the European Aquatics Championships in Bologna, Italy. Borg, known as the Swedish Sturgeon, scooped three gold medals in freestyle, breaking the record for the 1500 metres just a few hours after he'd had his front teeth knocked out playing water polo against France. He became a poster boy for the zippy Racerback costume, which he sported in ads promising 'the minimum of resistance' and 'maximum body exposure to tonic sunshine'.

Borg's suit was made by a Sydney company run by Scottish migrant Alexander MacRae. The winning entry in a staff competition gave the company the slogan 'Speed on in your Speedos'. MacRae renamed his swimwear division Speedo. By 1929, Speedo Racerback swimsuits were being produced for the public, and the company had become associated with the Olympics.

The journey towards a briefer Speedo style was well under way by the 1956 Melbourne Olympic Games, where Australian swimmers, including Murray Rose, wore fast-drying nylon versions that might have made Borg blush. Long gone was the full-body suit of the 1920s; the new costumes were hip-hugging and pared-back, although they were not yet the full budgie smuggler. (A pair of Rose's Speedos from the Rome Olympics in 1960 fetched $1000 at auction in 2017.)

In 1959, artist Peter Travis tapped into increasingly relaxed mores by creating a brief swimsuit for Speedo to be worn on the hips rather than at waist height. 'The hips are stable. It isn't like tying something around the stomach,' Travis recounted in 2008. Reducing the amount of fabric on the sides was a matter of dynamics. 'If you lift your leg up at right angles, that is the shape of the way it is cut.'

WERE BUDGIES REALLY SCANDALOUS?

Once it became available for sale in 1961, Travis's Speedo design quickly attracted the attention of Bondi Beach inspector Aub Laidlaw. Laidlaw wore a white Panama hat with 'Inspector' embroidered on the band and had a reputation for carrying a tape measure to assess the modesty of women's bikinis. If they failed to measure up, the wearer would be asked to leave the beach. One day, upon seeing the new Speedo design, Laidlaw summoned the police, who arrested the wearers for indecent exposure. Charges were dismissed, however, as the garments did not display pubic hair, thus protecting the decency of Bondi—and pushing sales upwards.

'The men of Sydney ... organised beach protests at Coogee and Manly, with the largest at Bondi.'

For Inger Sheil, assistant curator at the Australian National Maritime Museum in Sydney, accusations of indecency levelled against the swimsuit echo earlier outbursts of prudery on Australian beaches. 'In 1907, there was a proposal to the law that [male] bathers should wear neck-to-knee suits with modesty skirts,' Sheil says. 'Within two days, the men of Sydney rose up and organised beach protests at Coogee and Manly, with the largest at Bondi. They had a dead seagull attached to a banner and wore anything that they could cobble together to approximate women's clothing. It was a carnival atmosphere and the local councils recognised that it was an excessive law. Contested bodies dominate the history of swimwear in Australia.'

Swimming briefs may be popular on Australian beaches, but they're compulsory in many French pools. 'It's a question of hygiene,' says one French pool guide. 'Swimming trunks, as well as Bermuda shorts or board shorts, can be worn all day long. By banning this swimwear, the aim is above all to reduce pool pollution (hair, sweat, urine residue, etc.) to preserve water quality.' The pockets of shorts can also harbour nasties such as tissues, which have no business being in pools, the guide points out.

Meanwhile, in the United States, outside of charity runs and comedy sketches, swimming briefs remain an oddity. 'If only Freud could have lived long enough to dissect the semiotics of Speedos,' fashion commentator Simon Doonan told *Slate*. 'What would he have made of the US male's horror of being caught in a tiny swimsuit? . . . Speedo-wearing is also a cultural flashpoint. Revealing men's swim garments are, for the US consumer, irrevocably associated with "foreigners" and, most terrifying of all, friends of Dorothy.'

ARE BUDGIE SMUGGLERS GAY?

Images of swimming briefs are popular with the gay community for reasons that become increasingly obvious as the icy waters of winter warm up for summer. As homosexuality became more accepted in the 1990s and early 2000s, models wearing budgie smugglers featured on the covers of Australian magazines targeted at a gay audience, such as *Outrage* and *DNA*, inspiring a new breed of swimwear brands.

The aussieBum brand successfully garnered a gay following by aligning itself with gay political causes and sporting associations. The appeal accelerated when in 2003 a video clip for Kylie Minogue's hit single *Slow* featured swimmers wearing aussieBum costumes, attracting the attention of Selfridge's department store in London.

Another brand, Budgy Smuggler, features prints inspired by McDonald's, XXXX beer, bin chickens (long-billed ibises) and the Aboriginal flag on its swimming briefs. Jake Smith, the founder of the Australian swimwear label Smithers, which has supported LGBTQIA+ awareness with its Pride collections, says Budgy Smuggler has broadened the audience for swimming-brief styles. 'Men here [in Australia] are more adventurous and playful now,' says Smith.

So, budgie smugglers are not gay—they're pieces of fabric—but if you were going to impose a sexuality on them, fluid is the safest bet.

WHO WEARS BUDGIES IN POPULAR CULTURE?

Tony Abbott. And he isn't the only politician who has courted the media in Speedos. Scott Morrison and both Malcolms (Fraser and Turnbull) have sported the briefer bathers too. They've become part of an unofficial political uniform, along with Akubras in country areas and tracksuits for morning jogs. Bob Hawke was a fan, snapped in budgies many times over the years, including at an ALP conference (it was the '70s) and on holiday wearing just striped budgies and a watch in the '90s.

Richard Taylor, the curator of the *Bondi: A Biography* exhibition at the Museum of Sydney in 2011, summed up the swimwear's instant appeal to voters: 'There's certainly a common theme on Australian beaches that when you're stripped down to your Speedos, everybody looks the same. It's very hard to tell rich or poor, so when you're on the beach, everyone is kind of equal.' It's why Gough Whitlam allowed himself to be photographed in 1974 at the Eastland Shopping Centre in suburban Melbourne buying bright orange-and-pink Speedos for $1.50 (size 40). There is no photographic evidence of him wearing them, however.

'When you're stripped down to your Speedos, everyone is kind of equal.'

Budgies, or 'scungies', come under fire in the 1979 surf novel *Puberty Blues,* set in Sydney's Sutherland Shire: 'The ultimate disgrace for a surfie was to be seen in his scungies. They were too much like underpants.' Yet the book's co-author, Kathy Lette, has since celebrated the briefs, advising her British readers in *The Times*, 'Speedos, aka budgie smugglers, are so skimpy that you can detect a man's religion when he's wearing them.' Budgies do have a firm association with sporting prowess, from the 1984 movie *Coolangatta Gold*—in which Colin Friels and Joss McWilliams attempted to channel real-life Ironman champion Grant Kenny—to the surf-lifesaving reality show *Bondi Rescue.*

The appeal is global. Actor Jude Law attracted headlines with a swimsuit power move, wearing luminous white briefs for the 2016 television drama *The Young Pope.* But for the ultimate budgie appearance on celluloid, many critics reserve their thumbs up for Ray Winstone's gold trunks in *Sexy Beast* (2000). Winstone's retired safe-cracker Gary Dove brandishes his budgies as he sunbakes on a lounge in the south of Spain and reflects on how hot he is. 'Swelterin'.' The actor reportedly swiped the trunks from the set and continued to wear them until they became threadbare.

DO BUDGIES STILL CUT IT?

In the 1970s and '80s, images of toned athletes fresh from the Olympic pool helped fuel sales of Speedos, but with the launch of the controversial Fastskin suits in 2000 and the increasing popularity of bike-short-style 'jammers', budgies have faced stiff competition.

'It's still one of our most popular silhouettes in terms of volume,' says Speedo national sales manager Matthew White. 'In recent years, there has been a renaissance among twenty- and thirty-somethings, who are happy wearing it at the beach instead of a board-short style. It's a combination of a "taking the piss" element, where groups of guys wear lairy styles, and increased body confidence in customers.'

Swimwear designer and stylist Michael Azzollini says photos of men wearing budgies are shared on Instagram. 'I will often follow up on who is buying the brief styles,' he tells us. 'They are extremely body conscious and proud of their six-packs. A lot of the customers are gay, but there are also some proud straight guys.'

That body confidence fails to stretch to younger swimmers, though, who seem to take comfort in the greater coverage of the jammer style. Still, ultimately, the launch of Speedo jammers has failed to dent the traditional briefs' prime position, alongside aquashorts, which have extra centimetres at the hip but are a tween favourite. 'We are seeing six- to 14-year-olds wearing the jammers, perhaps to emulate their sporting heroes and also because they might feel more comfortable around their peers wearing them,' says White.

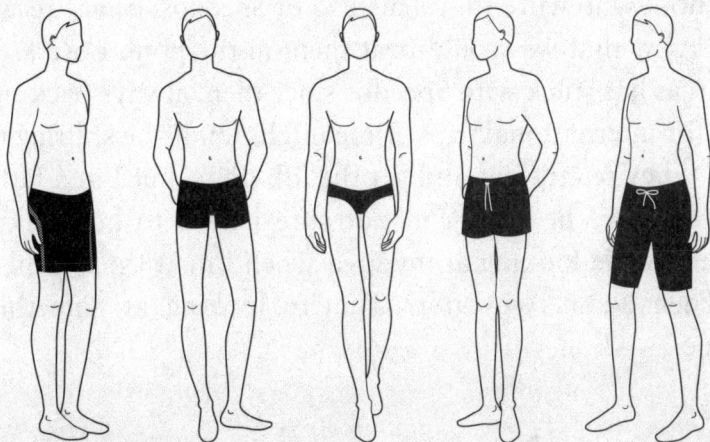

From left: Jammers, aquashorts, budgies, shorts style, boardies.
Simon Rattray

As for older customers, the brief styles still rank as the highest seller. 'There's a consumer that picks and sticks,' he says, referring to sun-pashed beach elders.

White says that even if you don't see swimming briefs at the local pool or beach, they are often lurking underneath other styles. 'A lot of Aussie males wear them beneath their board shorts or their wetsuits. The fact remains, when you're in the water swimming or bodysurfing, they are effective and don't drag.'

'They're called jammers for a reason,' says recreational swimmer Colin Standen, who trains with the Sydney club Wett Ones. 'They're not very comfortable and jam everything in to reduce drag for Olympic-level athletes. If you're not at the Olympics, traditional styles are best. The only drag that happens is when the elastic or fabric starts to give after too many washes. Then you can suffer from the saggy bum look.'

For caterer and podcaster Savva Savas, the front rather than the back is the issue. 'As a kid, I would laugh at the boys at school who wore shorts instead of Speedos,' Savas recalls. 'We knew that we would beat them in the pool. Losers.'

Savas has stuck with Speedos since then, always packing a pair for international trips. Time might finally be shifting the tide. 'They're efficient and get the job done, but I'm a father of two now,' he says. 'I'm not sure I need to be thinking about people looking at my area when I'm at the beach.

'Then again, why not? They're looking at something classic.'

3

DOES HANDWRITING STILL MATTER?

We have become creatures of the keyboard.
Yet writing, once considered an ancient
'handicraft', is still worth doing well.

Angus Holland

When he was the Prince of Wales, Charles III famously used a fountain pen and black ink to scrawl letters to British government ministers about everything from the war in Iraq to the plight of the Patagonian toothfish. His notes were typed up before being sent but he would then annotate them in his spidery handwriting, earning them the name 'the black spider memos'. When David Bowie penned the lyrics to 'The Jean Genie' on a lined notepad, he floated careful little bubbles over his 'i's. Centuries earlier, Leonardo da Vinci had written in his neat Latin script from right to left and in mirror image, possibly, it is believed, to avoid smudges because he was a left-hander.

Everybody's handwriting is individual. If you plan to forge a will, dash off a ransom note or leave a nasty message on your neighbour's car windscreen, beware: your writing might reveal your identity. If you plan to leave notes for posterity, your handwriting can be a liability too. 'Do you have any idea what the hell I'm saying there?' Joe Biden asked as he puzzled over handwritten notes with his biographer in 2017.

On the upside, whatever we trouble to write by hand is, by its nature, a unique artefact. It might last for seconds en route to the recycling bin or endure for years, even centuries. Victoria's State Library now houses the letter that bushranger Ned Kelly dictated to a fellow gang member over several months in 1879—'Dear Sir, I wish to acquaint you with some of the occurrences of the present past and future.' Far further back, the Herculaneum papyri, handwritten scrolls carbonised (and thus preserved, after a fashion) when nearby Mount Vesuvius erupted in AD 79, are finally being decoded, thanks to 3D X-rays and technology that uses artificial intelligence.

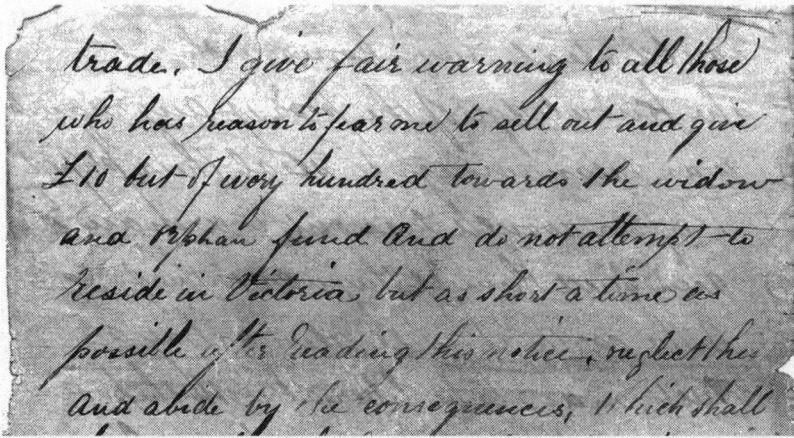

Joe Byrne's script in the Jerilderie letter. *State Library of Victoria*

Yet today we rely mostly on keyboards, tapping out our innermost thoughts as lines of digital typography, firing off emails and text messages with just the odd emoji where once we might have added an inky flourish. While we still learn to write—school exam answers are among the few things that must still be handwritten—the power of the pen is in decline. Or is it?

Is handwriting still important in the digital age? Does it have advantages over typing words? And can you really tell anything about somebody's personality from the way they write?

The power of the pen is in decline. Or is it?

IS HANDWRITING A WINDOW TO THE SOUL?

It has long been thought that handwriting offers insights into personality. So-called graphologists or graphoanalysts have claimed that the way we form our letters indicates whether we are outgoing, introverted or even deranged. In 1886, a Dr Peckham quoted in *The New York Times* declared that 'insanity' was 'readily distinguishable' in handwriting. 'The waviness of the lines, the size of the letters, the tremulous upstrokes are each eloquent of unhealthful tissue in a particular corner of the machinery of the mind.'

In 1939, New York psychologist Nadya Olyanova examined Adolf Hitler's handwriting and found 'three outstanding traits: indecision, depression and morbid introspection'. She continued: 'It takes no handwriting expert to recognise the cramped, drooping uncertain signature as a manifestation of the Fuhrer's cramped, self-centred approach to life.'

By the 1990s, US employers were using handwriting analysis to psychologically assess potential hires. 'Like dreams, handwriting is a visible expression of behavior that comes from the mind,' one graphoanalyst told *The New York Times*. 'It reveals our true personality—those inherent traits that cannot be forged.'

Even Bowie got the treatment when his papers came up for auction in 2016. One magazine had a handwriting expert look over the longhand lyrics to 'The Jean Genie' from 1972. 'For someone who was so experimental,' they noted, 'it is surprising that his writing is quite slow, upright to slightly left, with self-protective arcade structures . . .'

Can we really discern personality from pen strokes? Dr Carolyne Bird, a leading Australian forensic handwriting analyst, says this: 'There have been a number of studies looking at the claims graphologists make, in terms of what they can deduce [about personality] from handwritten forms, and the outcome of those is that they're no better than chance.'

That's not to say handwriting isn't interesting in its own way. How you hold a pen, how hard you press, where you start and in which direction you form letters (clockwise or otherwise)—all form a picture that is, if not a portal to the dark corners of your psyche, at least particular to you. Which is how handwriting can be incriminating.

An idiosyncratic note left when toddler Charles Lindbergh Jnr (son of the aviator) was kidnapped in New Jersey in 1932, and several ransom notes that followed became key evidence against the man eventually executed for the child's murder. The 'Hitler Diaries', sold to periodicals around the world, turned out to have been forged between 1981

Ransom notes became key evidence against the man eventually executed for the child's murder.

and 1983, with handwriting that didn't match Hitler's (and in books that had the wrong kind of paper and bindings for the era).

A ransom note found in the Lindbergh home in 1932.

In Australia, says Carolyne Bird, handwriting analysis can come into play with forged documents such as wills, suicide notes that may not have been written by the deceased, in cases of anonymous threatening letters, and with notes from drug dealers that are to be used as evidence, 'What they call tick lists—who's ordered what drugs from them and what they might owe'. The devil's in the details: how a writer crosses their 't's or spaces their letters, or even the scratch marks made by the tiny housing around the ball on a ball-point pen as it pushes into a sheet of paper.

The work of handwriting analyst Cliff Hobden includes checking whether signatures are genuine or suspect and scrutinising differences in letter design, proportion, alignment and spacing in any given sample. Hobden, who works for private investigation firm Lyonswood, compares scrutinising handwriting to investigating a road accident. 'You look at the road from above, you've got a pair of skid marks. Those skid marks, if they go off the road at a particular angle, will clearly show what direction the car was travelling in.'

WHERE DID HANDWRITING COME FROM?

Writing was once largely the province of the elite or the people they employed, such as the mediaeval scribes who painstakingly copied the Bible onto parchment with quills of goose or swan feathers dipped in ink. Exactly when 'handwriting' began to appear as an everyday activity is a matter of debate and definition: Egyptologists might consider hieroglyphs to be handwriting. What we can say with confidence is that writing, as opposed to simple drawings, began to appear from 3200 BC, largely on clay tablets in Egypt, China and Mesopotamia, the last the birthplace of cuneiform, a script that became the backbone of ancient written languages from Akkadian to Palaic.

An illustration of Babylonian cuneiform. *Thomas Faull/iStockphoto*

'This is a super-interesting point in human history where people say, "Okay, we need to write things down",' says Louise Pryke, a researcher at the University of Sydney and one of the few Australian scholars who can translate cuneiform. 'Trade becomes more complex, and social interactions and so forth become more complex. Administration is a driving force behind developing the writing system.'

The world's first-known named author, the Sumerian priestess Enheduanna, wrote in cuneiform around 2200 BC, notes Pryke. Back then, it wasn't uncommon for women to write—it was considered a handicraft. Cuneiform doesn't use an alphabet; instead, it is a collection of marks that look rather like 'chicken scratchings', says Pryke. The Phoenicians developed the first alphabet, about 1000 years later, which, oddly, had 22 consonants but no vowels.

Writing systems developed in many pockets of the globe, from Arabic scripts in the Middle East to Chinese *hanzi*, Japanese *kanji* and more. The Latin alphabet was first used in early Rome, evolving into rotund-looking 'uncial' scripts, set down on vellum (animal skin) by scribes. The oldest surviving handwritten documents from Britain are waxed wooden tablets (a bit like an ancient Etch A Sketch) made during the Roman occupation and unearthed from beneath

12 metres of river mud during excavations in London between 2010 and 2013. (The jottings included supply orders and complaints about the locals and the roads, which, as we know, the Romans eventually fixed.)

During the Renaissance, Venetian writing master Giovanni Antonio Tagliente offered a DIY approach to handwriting in his popular 1524 book *The True Art of Excellent Writing*. Because diplomatic scribes were among Tagliente's target audience, the book covered various scripts, including Hebrew and Arabic. Tagliente also published self-help books on arithmetic, embroidery and love letters (which, given the personal nature of handwriting, are surely still worth cracking out a pen for).

In the early 1800s, penmanship was a mark of status and class.

During the brief flowering of Regency England in the early 1800s, penmanship was a mark of status and class. Jane Austen's *Pride and Prejudice* is full of letter writers and the occasional handwriting snob. 'Oh!' exclaims Caroline Bingley of her brother's missives. 'Charles writes in the most careless way imaginable. He leaves out half his words and blots the rest.' Most educated folk had their own pen knife, a tool for sharpening their pens, which were still quills made from feathers dipped in ink.

The 'peak handwriting' era soon followed. By 1850, cursive was being taught in Australian schools, particularly forms of copperplate, an ornate style that emerged in the 1600s and that, over time, came to be engraved on to copper plates used in printing. Your grandparents probably had excellent handwriting; their grandparents almost certainly did. Ned Kelly actually dictated his Jerilderie letter—8000 words of flowing, looping cursive—to fellow gang member

Joe Byrne, who apparently had better handwriting, although Ned could write too.

By the 1970s, ornate variants had been largely replaced in schools by writing styles that were easier to learn and faster to use. There are currently five main types of joined-up handwriting taught in Australia, all derived from Modern Cursive, a style imported from Britain in the mid-1980s, according to Kevin Brown, whose website Australian School Fonts offers resources for students and educators. 'This new method prioritised cursive writing from the outset by making sure that the students were taught to form letters in such a way that it led naturally to cursive writing,' Brown says.

DO KIDS NEED TO LEARN HANDWRITING ANYMORE?

The death of handwriting has been a long time coming. Johannes Gutenberg was blamed for its decline when he invented the printing press in the mid-1400s, putting no doubt furious Bible-copying monks out of work. But handwriting persisted through the eras of the feather quill; then the fountain pen, with ink that flowed from internal reservoirs; followed by the ballpoint, popularised by Hungarian journalist Laszlo Biro in the late 1930s (although it was first patented in 1888). Typewriters, which marked a departure from hands-on writing, were commonplace for much of last century, with *The Sydney Morning Herald* describing them in 1952 as 'coming to the rescue of the lazy and the shamefaced'.

But the true extinction moment seems to be the arrival of laptops and tablets in schools and universities from the early 2000s, and not just in the West. Perfect hand script is becoming increasingly rare in China, reported *The South China Morning Post* in 2022. 'So much of communication in China and the world in general is digital that it is normal for the average person to go long periods without using traditional handwritten communications.'

Finland ditched cursive and offered more typing classes.

The necessity of teaching cursive is continually debated in educational circles, with battle lines drawn between its relevance in a world of keyboards versus its supposed benefits for early learning. In the United States, individual states have either abandoned or re-enforced the teaching of longhand. (Indiana dropped the requirement in 2011; California reinstated it in 2024.) 'For many students, cursive is becoming as foreign as ancient Egyptian hieroglyphics,' declared *The Washington Post* in 2013.

Finland, that educational Mecca, ditched cursive and offered more typing classes in 2016. Neighbouring Sweden swung in the other direction in 2023 when it tweaked its syllabus to give children more time to learn handwriting and less time on tablets. Learning to write by hand remains mandatory in Australian schools, but its teaching—and, critically, ongoing practice—is not as ruthlessly policed as it would have been a few decades ago. It might not be seen to matter, except that our students—with rare exceptions—are still required to complete exams by hand.

At least one school has owned up to the problem: St Andrew's Cathedral School in Sydney offers students out-of-hours coaching on posture, hand-strengthening and mobility. 'In recent years, students worldwide have faced increasing difficulty in maintaining legible handwriting at the required speed and endurance for formal examinations,' the school's deputy head of quality teaching, Dr Kirsten Macaulay, has said.

Children really need to learn to write properly and then to use it or lose it, says Mel Micallef, an occupational therapist in Melbourne. 'Often the legibility and all those foundational skills fall by the wayside and you can't read what they're writing,' she adds. During the pandemic, Micallef observed her own daughter, in prep, remote learning at the kitchen bench and wondered about all the other children who weren't getting intensive handwriting instruction 'and what the consequences would be down the track'. Micallef quit her job and started a business, Ready, Set, Write, helping young children who struggle with handwriting and assessing older students approaching exams who might need special consideration because their writing is so illegible it's deemed a handicap.

'Here are some words from older kids,' she says, pointing to her screen during our interview. 'What do you think that word is? Is it "lake"? No, "take".' We peer at another word. 'It looks like a Y but could be a funny-looking V or an incomplete H,' she says. We take a guess at a few more. Maybe 'video'? Yes, 'because of the E and O, you can sort of guess. A few illegible letters can change the whole word, making it impossible to read.' What is clear is that these students have put enormous effort into being understood yet even the most diligent examiner reading their essays would have to throw in the towel.

Unfortunately, chronic problems such as incorrect pen grip (clenched in a fist or between fingers rather than the classic 'tripod' between thumb, forefinger and middle finger) can take a lot of effort to remedy, says Micallef. 'It is really tricky to change pencil grip beyond Grade 2.' That said, just practising handwriting daily will help yours improve: she suggests dedicating 16 to 17 minutes a day to writing of any kind (that's around 100 hours a year). Micallef and other experts also suggest making sure the first and last letter of every word is always legible—with them in place, it's much easier to guess what the word is.

WHY BOTHER TO WRITE BY HAND?

There is plenty of evidence that writing by hand reaches parts of the brain that typing can't; that the unique combination of fine motor skills and thinking helps us process and retain information better. Studies have shown that handwriting can boost connections across brain regions, improving spelling accuracy, memory and conceptual understanding.

Writing by hand reaches parts of the brain that typing can't.

'When writing by hand, brain connectivity patterns are far more elaborate than when typewriting on a keyboard,' writes Professor Audrey van der Meer, from the Norwegian University of Science and Technology, in the journal *Frontiers in Psychology*. 'Such widespread brain connectivity is known to be crucial for memory formation and for encoding new information and, therefore, is beneficial for learning.'

When you take notes by hand, for example, you are more likely to consider what to write down, editing and condensing as you go, while bashing away at a keyboard can be more automatic. Indeed, a 2014 study by psychologists at the University of California found that students who took notes by hand understood the material better than those who used laptops. Similarly, in 2020, an analysis of 3000 US college students found their performance suffered when they typed their notes rather than writing them by hand. 'It's partly to do with information processing—it takes you longer to think about what you're going to write,' says Dr Annie McCluskey, an occupational therapist at the University of Sydney.

Several prominent novelists agree. Horror writer Stephen King, who wrote his novel *Dreamcatcher* with a fountain pen, once told an interviewer the fundamental nature of taking something in your fist and physically making letters on a page 'takes you back to childhood, when you learned to write and when you dared to dream'.

Science-fiction author Neal Stephenson uses a fountain pen despite his love of technology and future imaginings. Novelist Donna Tartt writes first in ballpoint pen, then uses colour-coded pencils for revisions. J.K. Rowling wrote the Harry Potter novels in ballpoint (her manuscripts later sold for hundreds of thousands of dollars). David Foster Wallace, who wrote the door-stopper *Infinite Jest*, once observed that while he could type much faster than write, 'writing makes me slow down in a way that helps me pay attention'.

In Australia, thriller writer Garry Disher has his own superstition about writing in longhand. 'I've heard other writers say the same thing, that there has to be the right set-up or all the magic leaks away,' he tells us. 'For me, it has to be blue biro, never black.' He handwrites a first draft then edits it as he types up the manuscript. 'I can write very quickly,' he says. 'I edit as I go along, I cross things out. I make notes to myself, draw arrows across the page.'

Melbourne poet and novelist Philip Salom takes a mixed approach: he writes prose on a computer but poetry in longhand. 'I have to write by hand when I'm writing poems,' he says, 'because it's sensuous.'

4

IS THE MOON FOR SALE?

A new era of lunar missions has begun. Who can stake a claim on Earth's closest neighbour?

Sherryn Groch and Felicity Lewis

US flags, footprints, broken spacecraft, some pluto-
nium, a couple of golf balls, a family snapshot, a
Bible and a surprisingly large amount of human
faeces. This is what humanity has left behind on the Moon.

In 1969, astronaut Neil Armstrong took that first 'giant
leap for mankind' onto the lunar surface, and 11 more
astronauts across five Apollo missions followed, collecting
rocks that changed our understanding of how the Earth and
the Moon came to be (and even, in 1971, sneaking in some
golf). But humans haven't been back since '72.

Now another space race is kicking off, this time driven
not only by governments but by a growing private sector
(and some enthusiastic billionaires). NASA plans to land
people on the Moon again in the next couple of years, the
first step to developing a permanent lunar base (with
the aid of an Australian-made rover). China, Russia and
other powers have similar ambitions. The Moon is seen as
a crucial launch pad for human missions to Mars, but it's
rich in resources of its own, from rare metals to 'ice water'.
Consequently, the Moon is a new frontier where space rules
will be made and tested.

Of course, scientists will tell you the Moon is more
important than any of that. Without its pull, life on Earth
may not even be possible to begin with. But, if you ask a
guy named Dennis Hope, who lives in California, *he* owns
the Moon—having filed the paperwork at his local council
40 years ago—and all these big lofty plans to develop it will
need his sign-off.

So how are space lawyers unravelling who gets to take
what from the Moon? What will the rules up there look
like? And what happens if not everyone agrees?

WHAT IS THE MOON, ANYWAY?

Getting our Moon was probably the best and the worst thing to ever happen to the Earth. The best because, without the stabilising gravity of our unusually large Moon, we wouldn't have the tides and the Earth would wobble on its axis, causing our climate to swing from searing hot to freezing cold, perhaps too quickly for life to adapt. The worst because it's thought our planet and another named Theia collided in the early solar system some 4.5 billion years ago, engulfing Earth in fire. The Moon is debris from this cataclysm, locked in Earth's orbit.

This kind of cosmic 'fender bender', as astrophysicist Professor Jonti Horner at the University of Queensland calls it, is the leading explanation for why our Moon is so big (about a quarter the size of the Earth) compared to other moons we've seen and why it's covered in a glittery rock (the cooled remains of a magma ocean).

The Moon is the brightest thing in our night sky, reflecting the Sun's light. Because it has no climate or ocean, no weathering of its rocks, it keeps a pristine geological record. 'Even with the naked eye, the Moon sometimes seems close enough to reach out and touch,' says NASA scientist Professor Darby Dyar. It *is* close, only 1.3 light seconds (or roughly 380,000 kilometres) from Earth, although each year it creeps four centimetres further away. Dyar watched the Apollo astronauts 'bouncing around [the Moon] on grainy television broadcasts' as a kid and in 1979, as a 21-year-old PhD student, she found herself studying lunar samples they brought home. 'You can understand why my hands shook every time I had to handle them,' she says. 'They still do!'

Those rocks changed our understanding of where the Moon came from. But they threw up new mysteries too.

Scientists thought the Moon must be a leftover chunk of Theia, likely its old core. But instead of heavy elements such as iron that we'd expect to find inside an old, rocky planet such as Theia, the Moon's insides are lighter, more like Earth's mantle and crust, with a tiny core. 'Every planet bears the scars of how they formed,' Horner says. Chemically, 'it looks like the Moon and Earth formed in the same place'.

The radio waves that humans have been beaming into space since the invention of broadcasting don't reach around to the far side of the Moon. That's why China landing a spacecraft there for the first time in 2019, using a relay satellite, was such a feat. 'We are exploring more of the Moon now,' Horner says. 'Who knows what we'll find next?'

In 2020, China collected the first fresh samples of Moon rocks in decades and, in 2024, the first soil from the Moon's deepest crater at the south pole. That crater is so large it's thought the meteorite impact that made it might have punctured the Moon's crust and exposed part of the upper mantle, says Dyar, who will lead one of several studies when NASA collects Moon rocks.

One huge find in 2018 was NASA's confirmation that ice water, detected for years by lunar missions (notably by India's), was not confined to cold, dark craters but was spread across the Moon's sunlit surface as well. The discovery suddenly made frequent trips to the Moon and even lunar bases much more feasible, just as rockets were plummeting in price, thanks to lighter and more reusable designs developed by a new generation of space companies

The discovery suddenly made frequent trips to the Moon and even lunar bases much more feasible.

including billionaires Elon Musk's SpaceX and Jeff Bezos' Blue Origin. Launching water into space is expensive—every kilogram requires more rocket fuel to escape Earth's powerful gravity.

Water can be split into its atomic parts: hydrogen and oxygen. Oxygen is obviously handy for life on an enclosed Moon base, while hydrogen is a component of rocket fuel, says Horner. This is one reason that NASA chose the Moon's south pole as the site of its next landing: the ice. 'That crater is loaded with it,' says Horner. The European Space Agency has also unveiled plans to mine lunar ice and dust in the coming years; China and Russia have their own plans to establish a joint lunar research station at the south pole in the 2030s.

There have been some big dollar estimates for the value of lunar water mining. Silicon Valley space start-up Moon Express has called lunar water 'the oil of the solar system' and the Moon 'the eighth continent'. But Dyar stresses that water on the Moon is 'not a renewable resource'. The exact amount there is unclear. Some imagine the Moon will instead become our 'gas station' in space, Horner explains, an ideal base to refuel spacecraft as we mine nearby asteroids rich in rare metals or head off on longer missions, including to Mars and Jupiter's icy moons.

In 2023, a NASA spacecraft set off on a six-year trip to Psyche, an asteroid so rich in metals, including nickel and gold, it's said to be worth more than Earth's entire economy. Our Moon has rare metals of its own, the kind used in electronics and green energy technology. Mining these metals on Earth is difficult and environmentally damaging because they are often deep underground. 'A lot of things that are not very rare in the cosmos are *very* rare in the crust of the Earth,' says Horner.

But there's something else on the Moon some experts think could turn it into the next Persian Gulf: helium-3. Given the Moon has virtually no atmosphere, particles from the sun have rained down on it for eons, including this rare isotope. Helium-3 can be used in nuclear fusion but isn't radioactive, making it a potentially huge source of energy. On Earth, it's so rare that it has an astronomical price tag. (China, in particular, has been mapping estimated reserves on the Moon.)

Of course, it's not all about stripping the Moon for parts. There are hopes that solar panels on its surface could beam huge quantities of renewable energy to Earth. The Moon also has very low gravity—just one-sixth that of Earth. Everything is lighter, including heavy machinery, and companies such as Blue Origin are eyeing it as a new manufacturing hub.

The Moon's lower gravity explains in part why clocks run about 58 microseconds faster there than on Earth. But when visitors start arriving on the Moon from so many countries and different time zones, 'Houston time', which was used during the Apollo program, won't cut it. The White House has asked NASA to establish a Coordinated Lunar Time as a 'stable reference point' by 2026.

SO, WHO OWNS THE MOON?

Dennis Hope says he has owned the Moon for more than 40 years. He was dreaming of getting on the property ladder when he looked up at a full Moon one night and thought, *Now, there's a lot of real estate.* Then he remembered learning about the Outer Space Treaty—the 1967 agreement designed to keep the peace in space. Signed by more than 100 nations, the treaty rules out any national appropriation of the Moon and other celestial bodies (or installing nuclear weapons on them, for that matter). But it doesn't explicitly prohibit individuals from owning land off-world.

Dennis Hope says he has owned the Moon for more than 40 years.

Hope quickly filed a notice of ownership for the Moon (and the other planets in the solar system) at his local council and in 1980 set up the Lunar Embassy, a business selling small plots of lunar land for about US$25 (as well as acres on Mars, Jupiter and beyond). The company has sold land to more than 6.5 million people, says Hope's son Chris Lamar, now CEO of the Lunar Embassy. 'We've had celebrities, actors, former presidents buy.' Although Hope wrote to the United Nations and the two major space powers at the time—the United States and the Soviet Union—asking if his claim was illegal, he has yet to hear back from any of them.

The company is built on the fact that many countries have laws allowing people to claim unoccupied land, and space is the ultimate uncharted territory. 'No one else was doing it and we got there first,' Lamar says.

Legal experts point out that Hope's 'loophole' wouldn't hold up in any court. The Outer Space Treaty, created during the space race at the height of the Cold War, established the Moon (and everything in space) as a global commons, a bit like the high seas. 'We all have a stake in it,' explains Steven Freeland, Emeritus Professor at the Western Sydney University School of Law and Bond University. 'By all rights, it would have been great if they had also put up a UN flag' when the United States first landed on the Moon in 1969. 'That US flag was an expression of national pride and prestige, but even then, America rushed to reassure the UN they hadn't really claimed the Moon with it.'

Under the treaty, nations are responsible for ensuring their citizens follow the spirit of international law, which rules out private ownership of land in space. (Freeland's missing out too. One of his classes bought him a Lunar Embassy plot 'with all the title deeds', he says. 'It's a novelty item.') Still, the treaty left room for interpretation. In 1967, the United States and Russia knew they were the big players, Freeland explains, and set out some rules to work together—and to reassure the international community about their activities—but 'why would they bind their own hands?'.

A decade later, as the potential for resource exploitation first came into focus, the UN developed a new Moon Agreement. The Moon and its natural resources were the common heritage of mankind, it said, but exploiting those resources shouldn't disrupt the Moon's environment and the benefits should be shared equitably (including among developing countries). An international regime would also be needed to manage lunar mining. Yet, when push came to shove, only 18 countries agreed to be bound by the Moon Agreement and in 2023 Saudi Arabia dropped out. Back in 1979, US lobbyists had argued it would create 'a communistic approach to space', says Freeland. (Australia ratified the agreement in 1986 under Prime Minister Bob Hawke, who also marshalled support for the Madrid Protocol of 1991, which prohibits mining in Antarctica.)

Now experts such as Freeland are finding a legal way forward. He co-chairs a UN working group charged with examining laws on space resource exploitation. It has managed to get consensus among more than 100 nations, including Russia, the United States and China, on how to proceed, even in the midst of Russia's war in Ukraine. Whether this leads to any new rules or treaty remains to be seen. What's clear, says Freeland, is that nations see big opportunities on the Moon, including in science.

Already, a new wave of robotic missions are underway, with the first landing by a private company, US group Intuitive Machines, happening in February 2024. Australia, Canada, Europe, Japan and the United Arab Emirates have all joined NASA's Artemis program. Australia is developing the suitcase-sized 'Roo-ver', which will collect lunar soil to see if oxygen can be extracted from it, and Australian scientists are involved in a project to try to grow crops in space. As China rises as a new force in space exploration, it is collaborating with Russia in establishing a Moon presence, says Dr Malcolm Davis at the Australian Strategic Policy Institute.

Australian scientists are involved in a project to grow crops in space.

While NASA continues to work with Russia on the International Space Station, which will be decommissioned in a few years, US law bans it from collaborating with China unless it has special authorisation from US Congress or the FBI.

With human habitation likely in the next 20 years, questions arise as to what best behaviour would look like on the Moon. For one thing, lunar bases could give rise to some bendy notions of ownership, says University of Adelaide professor Dale Stephens, who co-edited the first comprehensive 'user's guide' to military laws in outer space, *The Woomera Manual*, covering both peacetime and rules of armed conflict. Under the Outer Space Treaty, he says, visitors can enter your living facilities in space, so long as you agree. In other words, you can say no. 'This is the closest you're going to get to a level of exclusivity as to your facilities,' says Stephens. 'Your little facility effectively will be your "territory" because, on the basis of reciprocity, you can restrict others from coming to your habitat.'

The Artemis Accords, agreements first signed by NASA and eight other space agencies (including Australia's) in 2020, set out a code of conduct for sustainable lunar development. Russia and China haven't signed on, claiming the accords are designed to protect America's edge in space. But by mid-2024, 39 countries were on board. 'They're voting with their feet,' says Stephens. Even these accords, although not binding international law, offer opportunities for quasi-control, he notes. Parties can declare temporary safety zones around potentially harmful activities, such as mining, and request (but not demand) that others keep out. 'I think that this idea of safety zones is going to be the way that groups create a peaceful order while going about their business on the Moon.'

Head of NASA Bill Nelson ruffled feathers in 2022 when he warned that China may be looking to claim the Moon, possibly by a creeping territory grab. 'We must be very concerned that China is landing on the Moon and saying, "It's ours now and you stay out",' he told a German newspaper. China denounced the comment as a 'lie'.

To Freeland, nothing could be worse than countries operating on the Moon under different rules. 'That's a recipe for misunderstanding, miscalculations and worse,' he says. 'We have to find a way, that even if they're not co-operating with one another, they agree on and abide by the rules of the road.'

WHO WILL PROFIT FROM THE MOON?

If the first space race was all a matter of prestige, the new push to the Moon is largely about money. Government agencies are building partnerships with private interests—so-called 'new space' companies, says Malcolm Davis. 'Although the Moon is central to a lot of countries' plans, it's more about commercial gain now [as well as] national security and defence.' In 2020, Morgan Stanley estimated the global space industry could generate more than $1 trillion of revenue in 2040, at least half of it from satellite broadband services.

Against this backdrop, NASA is contracting companies to bring Moon rocks to Earth for small payments. NASA could do this itself but, experts say, the agency has instead decided to establish, for the first time, a market to buy and sell things from the Moon. And crucially, companies such as SpaceX and Blue Origin are helping supply cheaper, lighter rockets. (Even NASA's Artemis missions will use SpaceX rockets.)

Lobbying by companies led to the United States passing the *Space Act* in 2015. The act interprets the Outer Space Treaty as allowing US citizens to own, use and sell resources they find in space. The idea has been compared to fishing in international waters—you might not own the water, but you can sell the fish. Luxembourg, Japan and the United Arab Emirates have similar laws that apply only to them. No one will spend money on space innovation otherwise, proponents argue, and too much regulation could kill the burgeoning commercial space sector. Without established rules, however, 'it's like the wild, wild West', says Davis. 'It's a free-for-all.' Experts worry that a gold rush will trump the Moon's potential use 'for the benefit of all', as the Moon Treaty puts it. Much of what is extracted in space is likely to stay off-world anyway to further interstellar expansion.

Without established rules, 'it's like the wild, wild West'.

When NASA lands the first woman and the first person of colour on the Moon this decade, it will 'inspire a new generation', Horner says. But he thinks the main benefit of the new space race will be in spin-off technologies on Earth as engineers and scientists rush to tackle space's challenges.

Of course, developing the Moon won't be easy. One challenge? Moon dust. Unlike weathered particles on Earth, Moon dust is sharp and poses serious risks to human lungs and equipment. And, without a protective atmosphere, Moon bases will be exposed to various perils, from radiation to rocket crashes and more trash. 'And if we screw up space . . . we'll all suffer because we all depend on space,' says Freeland. 'The big [countries], who are the most dependent on space, have the most to lose—and therefore the greatest incentive to ensure this doesn't happen.'

Can we wreck the Moon? Technically, we've already contaminated it. When the *Apollo* astronauts left bags of their poo to save weight on the trip home, they also dumped colonies of microscopic life. If something is still alive in it when humans return, we may have unintentionally proven that life can cross-pollinate between worlds, perhaps a crucial new twist in our quest to find aliens. Either way, rubbish will need to be one of the first issues ironed out when a lunar base is set up, says Stephens. On the plus side, Horner notes, for all our technological advancement, we at least don't have the means to knock the Moon out of its orbit or interrupt Earth's tides.

COULD WAR BREAK OUT ON THE MOON?

In the 2019 TV drama *For All Mankind*, a group of Russian cosmonauts launch an ambush on a US Moon base. The show reimagines history by asking an intriguing question: what if

Russia had landed on the Moon first? Yet violent lunar scenarios are already being contemplated by real-life diplomats, politicians and space lawyers. Space already plays a key role in war. In the 1960s, it was still 'a very niche thing for defence forces', says Davis. Now 'everyone plugs into space, and we can't fight without it'. Destroying or jamming satellites that relay military communications, launching cyberattacks, even taking over strategic points in space, including potentially on the Moon—these could all come into play, even though space is intended to be used peacefully.

> **Violent lunar scenarios are already being contemplated by diplomats, politicians and space lawyers.**

Donald Trump's administration was mocked for creating a new arm of the US military, Space Force, in 2019, but many countries incorporate smaller space units within their existing forces, including Russia and China. Although weapons testing is illegal on the Moon, carrying a weapon is not, unless it's one of mass destruction. Russian cosmonauts have reportedly carried guns in their landing survival kits. After all, in 1965, two returning cosmonauts were stranded for days in Siberia—bear country. While astronauts pulling guns to contest a lunar claim may seem quaint, Davis says it shouldn't be dismissed out of hand, even if he thinks most 'warfare in space will be done by robots and uncrewed weapons systems, satellites as opposed to X-wing fighters in *Star Wars*'.

Conflict could, for example, spread from Earth. 'If you've got two countries fighting it out on Earth,' says Stephens, 'it is possible that they may take that fight [to] outer space, including the Moon.' In that situation, laws on the use of force and armed conflict—such as responding in proportion to a threat—would stand, including the obligation to avoid harming non-combatants. 'If you've got soldiers fighting on the Moon, you've got a lot of civilians—science and mining, for example—then basic principles apply.' The prohibition on destroying infrastructure where it's indispensable to the survival of a population would take on new twists too. 'If you're up on the Moon, things like oxygen supply and water become quite critical,' Stephens says.

Whether technology matches the warp-speed dreams of industry or not, Freeland agrees that our history on Earth should be taken seriously. Wars are 'all about countries trying to claim territory or claim resources'.

Civilians in space—tourists, asteroid miners and so on, likely with varying degrees of training—will further complicate the picture. Astronauts are 'envoys of mankind', with an expectation under international law that they'll be rescued if they get into trouble. But 'in a settlement on the Moon, if you had a taikonaut, a cosmonaut, an Afronaut, [say] 20 people from 20 different nations all living together, all subject to their own national laws . . . what law applies?' Freeland asks. On the International Space Station, he says, such murkiness is sorted out ad hoc. The Moon is equally dangerous. 'I can't open a window because I'll kill everyone.'

Stephens hopes his manual, based on consultations with space-faring nations and various NGOs, will help clarify some of the legal ambiguities about space. Could the actions of a private operator on the Moon spark war, for example? 'We say that this is not possible,' Stephens says. 'A company can't unilaterally plunge Australia into an armed conflict. We took this position, in our draft manual, to a group of 24 states at the Hague and we think we are right.' The Outer Space Treaty says countries are always responsible for their citizens and their registered private companies 'but you can read that in a different way'.

He also thinks laws of war that ban trampling historic sites on enemy territory could extend to significant objects on the Moon. The base of the *Apollo 11* lunar lander remains in the dust with a plaque: 'We came in peace for all mankind'. And what of astronaut Charles Duke's 1972 photo of his family in flares, which he left in the lunar dust? Says Stephens: 'My view is that its global and historic significance may well entitle it to protection as cultural heritage on the Moon under the Artemis Accords and the 1954 Cultural Property Convention.'

Stephens believes we are entering a 'new golden age' both for humans in space and for the laws that apply there. At the moment, 'you've got the tension of some countries pushing, others pushing back. But then I think [there'll be] a general recognition by everybody: let's deal with it.' Davis is less optimistic, saying while Western countries may 'go forward with all these new treaties and [regulations], patting ourselves on the back', China and the Russia might just do their own thing. 'If all we're relying on is a hope that they have good intentions, well, hope is not a strategy.'

5

WHAT'S PICKLEBALL?

How did a backyard game invented by three dads go global—and how do you win?

Angus Dalton

The world's largest active volcano was erupting behind us but at ground level, it was the competitive spirit that was molten hot. I glared into my opponent's eyes. A moment earlier, he'd been my boyfriend with a holiday tan. Now, on this small concrete court beside a highway in Hawaii, he was simply my arch pickleball rival.

Looking for vacation fun, we'd picked up a set of paddles and a lime-green ball studded with holes, nestled between the rifles and spear guns on the shelves of a Walmart. If it had been tennis, a game I know my way around, my foe/boyfriend would have been toast. But in this strange new sport he was firing underarm serves with ease. We duelled in epic, rapid-fire volleys with a series of satisfying *thunks*. The volcano hurled glowing lava skyward and a smoky haze— a volcanic fog dubbed 'vog'—settled over the court.

Only later, as we sipped mai tais and the tension started to dissolve, did we realise we'd just indulged in America's fastest-growing sport, embraced by celebrities and professional athletes. Cricketer Steve Smith has invested in a Sydney-based professional pickleball team, for example, and tennis stars Nick Kyrgios and Naomi Osaka part-own a club in Miami. Back home after our stoush in Hawaii, we noticed that pickleball had taken hold on Australian shores.

So, what is this pickleball? Why is a game invented in Seattle in the 1960s taking hold in Australia? And, most importantly, how do you win?

WHERE DID PICKLEBALL COME FROM?

In the summer of 1965, businessman Joel Pritchard (who went on to become a US congressman) invited friend Bill Bell to stay at his summer home on Bainbridge Island, off Seattle in Washington State. After the pair returned from a game of golf, they were confronted by Pritchard's sulking 13-year-old son, Frank. 'I was bitching to my dad that there was nothing to do on Bainbridge,' Frank later recalled to *Pickleball Magazine*. 'He said that when they were kids, they'd make games up.'

Frank retorted at the time: 'Oh, really? Then why don't *you* go make up a game?'

The men took to the backyard badminton court with two ping-pong paddles, a perforated plastic ball from Frank's baseball practice gear and an inventive spirit. They recruited neighbour Barney McCallum—who was later instrumental in developing the game's rules and equipment—to make larger paddles. The game they pulled together from various sports reminded Pritchard's wife, Joan, of the friendly races she'd watched at university regattas. Oarsmen who had been dropped from the main competition would compete in 'pickle boats', an old maritime term for the last boat in a fleet of trawlers to return to port, charged with the pickling of the catch. Joan dubbed the new game 'pickleball'. (The Pritchards later owned a cockapoo called Pickles but, contrary to myth, the game was not named after their dog.)

The family built a proper pickleball court two years later and spread the word. In 1976, the first pickleball tournament was held in Washington State and by 1999 there were clubs in every US state.

Often described as a blend of tennis, badminton and ping-pong, pickleball also shares some features with chess, says Australian coach Ian Hutchinson. It's a game of strategy where players try to stay two or three shots ahead. 'It's also partly cricket, but only the sledging part,' he adds. The game's close quarters

A blend of tennis, badminton and ping-pong, pickleball also shares features with chess.

certainly encourage banter—it's played on a hard court about a quarter the size of a tennis court—and while it can be played as singles, doubles is more popular.

HOW DID PICKLEBALL TAKE OFF?

Since those early days in Seattle, a governing body for pickleball has emerged. The International Pickleball Federation, based in the town of Chevy Chase in the state of Maryland, had 77 member nations by 2024. Its goal: to get pickleball accepted as an Olympic sport.

Jen Ramamurthy, the Brisbane-based director of the Pickleball Australia Association, first picked up a paddle in 2018 after seeing a segment about the quirky sport on TV. 'When I started there were only about 10 people playing in Brisbane,' she says. 'Within that year there would've been thousands.' The association was founded in 2020 and membership surged to 4000 within two years. Today, Ramamurthy estimates, between 25,000 and 30,000 people play pickleball in Australia and there are 70 clubs and associations. Many people picked up pickleball during the COVID-19 pandemic when they could roll out a net on their driveway. 'It really skyrocketed. There's a phrase out there that pickleball is the most popular game no one's ever heard of.'

'Pickleball is the most popular game no one's ever heard of.'

Like most addictive exploits, pickleball is easy to pick up. Newbies can play a competitive game within about an hour, says Ramamurthy. 'You can have a nine-year-old, a 27-year-old, a 50-year-old and a 70-year-old on the court together and still have a good game. That's quite rare.' The annual Australian Pickleball Championships include hybrid events in which players with a disability, some in wheel-chairs, compete alongside non-disabled players.

The game has gone upscale in North America—it was the new golf at resorts from Miami to Mexico, according to *Architectural Digest* in 2022. It's also, unsurprisingly, popular on cruise ships. And a wave of trendy city businesses were bundling craft food, karaoke and cocktails with pickleball as developers set up courts in old warehouses, reported *The New York Times*.

Celebrities jumping on the bandwagon have added to the game's appeal. Bill Gates, who grew up in Seattle, has revealed he was in love with the game even before he started Microsoft in 1975. Former talk-show host Ellen DeGeneres has released a line of gorilla-themed pickleball paddles. And *The Late Show*'s Stephen Colbert hosted a pickleball tournament featuring pop musician Kelly Rowland, actor Emma Watson and Australia's Murray Bartlett, who won an Emmy for his star turn as a manager at a Hawaiian resort in the TV series *The White Lotus*.

Some stars are putting money into pickleball. Steve Smith became the first owner of a team in Australia's professional major league, which arrived in 2023, when he bought the rights to the Sydney Smash. 'I was drawn to pickleball for its incredible power to connect people,' he said. In the United States, basketball superstar LeBron James invested in the Major League Pickleball (MLP) competition after its debut in 2021. Kyrgios and Osaka and National Football League champion Patrick Mahomes are among the co-owners of the MLP's Miami Pickleball Club. Swimmer Michael Phelps, model Heidi Klum, basketballer Kevin Durant, tennis player Kim Clijsters, rapper Drake and actor Michael B. Jordan all own shares in other MPL clubs.

In 2022, a typo-laden Twitter spat broke out between Kygrios and former tennis doubles No. 1 Rennae Stubbs, who vowed never to watch pickleball, let alone invest in it. 'I would rather watch pain[t] dry,' Stubbs tweeted. 'Why all these tennis players think Pickleball is worth investing in & not the game that made them all the $$$ is beyond me.' Kyrgios fired back: 'I think LEBRON JAMES & KEVIN DURANT have a bit more of an idea [o]f what to invest in.'

But it's everyday retirees who are also a big driver of pickleball. 'There are retirement villages now that are building pickleball courts instead of bowling greens,' says Hutchinson. One luxury retirement developer in country Victoria advertises 'pickleball, billiards and late-night happy hours'. Pro and amateur competitions have both masters and open divisions. 'I'm 55 and I'm still playing in the open division,' says Ramamurthy, 'so I'm battling it out against 20- and 30-year-olds.

'It's not all about power and speed. There's a lot of brains, there's a lot of touch.'

It's not all about power and speed. There's a lot of brains, there's a lot of touch. It's been really good for us in terms of physical health, mental health, the friends we've met. It's so good for the elderly, getting them out of the house and playing sport again.'

Former primary-school teacher Sarah Burr is a pickleball star at 38. Before professional pickleball landed in Australia, she had to travel to the United States to compete for serious prize money. Now she captains Gold Coast Glory in the Pacific Pickleball league, which in 2024 offered $700,000 in total winnings, the largest cash prize pool outside the United States. 'Normally, when you're in your mid-thirties you're having to retire from a competitive sport, particularly professional sport,' says Burr. 'Whereas with this, you can be my age starting out, or you can be 50-plus starting out, and actually touring and making money.'

SO HOW DO YOU PLAY PICKLEBALL?

You'll need a small paddle and a plastic Wiffle ball (Wiffle ball is a scaled-back version of baseball invented in the 1950s). The court is 13.4 metres long and 6.1 metres wide (the same size as a doubles badminton court). The net is 91.4 centimetres high at the posts and 86 in the middle (five centimetres lower than the centre of a tennis net).

The ball must bounce within the baselines and sidelines but most red-hot pickleball action happens at the edges of a 'non-volley zone' either side of the net, known as the 'kitchen'. Play begins with an underarm cross-court serve from the baseline, which must clear the kitchen and land in the diagonal service box. The server gets one attempt and there are no lets. They must allow the return to bounce once on their side before hitting it. Then it's game on. Only the serving team can score a point. If the receivers let the ball bounce twice on their side, or hit the ball into the net or out of the court, the server wins a point. If they volley with a foot in the kitchen, a 'fault' is called and the point is lost.

While pickleball and tennis share some similarities, they are frenemies at best, particularly in the United States, where little-used tennis courts are being flipped for pickleball (four pickleball courts fit on one tennis court). 'There is definitely a little bit of backlash,' says Burr. 'Usually, the issues come if it's tennis players who are losing their facilities to pickleball courts.' Hutchinson believes the sports can co-exist. 'From a revenue perspective, I think tennis centres would be crazy if they didn't start to diversify at least some of their courts into multipurpose tennis and pickleball courts.'

Pickleball advocates say the sport's hand-eye coordination, positioning and volleying skills can help prepare kids for tennis and pickleball offers ageing or injured tennis players a way to stay active. But there are limits. Tennis types, dubbed 'bangers', are quickly put in their place, says Hutchinson. 'A lot of tennis players are still trying to imagine they're playing with a racquet with strings and a soft ball, and they're so used to doing topspin, whereas in pickleball you don't get much benefit from spinning the ball because it's a hard ball and hard paddle. With a tennis player banger, it's about trying to get them to slow the ball down. The better the [pickleball] player you get, funnily enough, the more finesse and the softer the game you play.'

Can pickleball undermine your tennis game, the way squash sometimes does? Sarah Burr, who hasn't played tennis in a while, says she'd struggle with tennis now that she's so enmeshed in pickleball's unique mix of shots and the slower pace you need to win. But she says people who play both games regularly can easily switch between the two. Pickleball can also affect other racquet sports. 'I picked up a badminton racquet about a month ago and I was hopeless because my brain's so calibrated to the pickleball paddle,' says Hutchinson.

BUT, COME ON, HOW DO YOU WIN?

Sets are played to at least 11 points, after which you must win by a margin of two points. Winning a point comes down to subtlety and strategy. 'The team that makes it forward to that kitchen line first and stays there is usually the team that wins the point,' says Burr.

Hutchinson notes that players are 'trying to manoeuvre people out of position and then wait for them to hit a loose shot or what's called the "pop up", where they get it up by mistake, which then allows the other team to hit it down for a winner through the middle.'

A dink—a drop shot into the kitchen—can be a formidable weapon. 'Doing drop shots into that seven-foot area is actually a very important strategic part of the game because that neutralises the opposition to be able to hit any attacking shots,' says Hutchinson. 'Because the ball is plastic, it doesn't bounce that high. So, dropping it over the net into the kitchen, it dies pretty quickly.'

Some trick shots are borrowed from tennis: between-the-leg 'tweeners', around-the-post winners. Others are unique to the sport. Players can leap across the kitchen for a smash and land off-court for a shot called an Erne (named for Erne Perry, the pro player who popularised the move). Leaping in front of your partner for the shot is called a Bert (as in *Sesame Street*).

'There's another really wacky shot that just happened recently, almost by accident, when a girl wrapped both hands around the paddle, hit [the ball] behind her head and landed this shot,' says Burr. 'Her name's Mary, so now that's called the Mary-go-round. There's literally just crazy stuff like this happening all the time.'

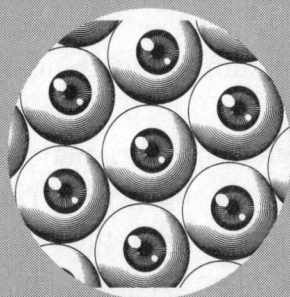

6

WHY IS PUBLIC SPEAKING SO SCARY?

Some people swear they'd rather skydive or bungee jump. Why does talking in front of others instil such terror— and can we learn to enjoy it?

Jackson Graham

Justin would do almost anything for his best mate. They'd been friends for 20 years, went to school and worked together. When his mate asked Justin if he could borrow cash for an engagement ring (so his girlfriend wouldn't spot the withdrawal in their bank account), Justin didn't hesitate. And when he asked Justin to be his best man at the wedding, he naturally said yes. But inside, he felt dread. Justin imagined making the obligatory speech: standing at a microphone in front of a room full of people, hushed and staring at him as his heart raced. It would be like every other time he'd had to stand up in front of people and speak. 'I just want to run away,' he says.

As a boy, Justin had confidence in spades. He gave school presentations, took on roles in plays and made speeches at birthday parties. But then he started work on a building site—and something changed. Every morning, he gathers with his co-workers for a meeting where they're all expected to speak up. When it's his turn, his adrenaline starts pumping and he feels lightheaded. 'I get it out as quickly as I can, just blankly, so it's finished and over,' says Justin, who asked us not to use his real name (for how much more nervous would he be knowing his audience had read about his nerves!).

Between 60 and 70 per cent of people experience moderate to high levels of fear about speaking in front of others.

Justin is far from alone. Studies say between 60 and 70 per cent of people experience moderate to high levels of fear about speaking in front of others. Some people swear they'd feel more relaxed parachuting or bungee jumping. Even people who speak publicly a lot—business figures, advocates, sportspeople and entertainers—can feel trepidation at the prospect. If we're all honest, says Queensland University of Technology lecturer Lesley Irvine, who has researched public speaking anxiety, there are going to be some situations where we all feel 'a degree of speaking anxiety'.

Why the fear? How can you overcome it? And what's the cost of not speaking up?

WHAT'S A PUBLIC SPEECH?

Before humans could write, we passed on stories and ideas through speech. Ancient Egyptian 'wisdom books' describe eloquence as the 'principle of fine speech', while Confucius said an 'artful tongue' was a gift 'one can hardly get on [without]'. In Greece, the power of persuasion was called rhetoric, an art form that could sway courts and politics. The use of rational speech was 'more distinctive of a human being than the use of his limbs', declared Aristotle.

Today, public speaking is still one way to be an active citizen, says Irvine. 'You get to talk through and present issues. You get to present your ideas. You get to advocate, whether that be in communities or at work.'

The term 'public speaking' may conjure images of spot-lights, lecterns and momentous speeches—the Gettysburg Address (Abraham Lincoln, 1863) or 'I have a dream' (Martin Luther King Jnr, 1963)—but in truth, it is every-where. It's the health and safety briefing from your colleague, the argument your neighbour raises at a community meeting, the lectures students listen to, the 'few words' said or the pre-recorded messages played at birthday parties.

'Public speaking is an extended speaking opportunity that can be formal or informal and be live, online, recorded or face-to-face,' says Irvine. 'When we say "extended", it doesn't have to be a very long time at all.' (Lincoln's much-lauded speech at Gettysburg lasted all of two minutes; it followed a two-hour address, now forgotten, by a former dean of Harvard University.)

A speech follows different rules than a conversation, says Matt Abrahams, a lecturer at Stanford University in California and author of the book *Think Faster, Talk Smarter*. In a speech, the speaker is essentially alone, without a wingman, no one to bounce off, all eyes on them as they stand in the spotlight. 'In a conversation, people are more actively and equally involved.' (Technically, he points out, a speaker can interact with their audience: 'I can take polls, I can ask questions'.) Speakers don't play characters and a speech isn't a performance as such, although it can have elements of drama and cause stage fright. 'If you're an introvert, you might need to lift your energy,' says speaking coach Sarah Denholm. 'That's certainly what I do; I'm being myself but a bigger version.'

WHY DO WE FEAR PUBLIC SPEAKING?

Sean McCaul was 10 when a teacher asked him to sing *Silent Night* at a school assembly. His older brother had acted in three school productions and his sister had written a play. No pressure, then. 'I remember singing and feeling all these eyes on me. I could feel my voice breaking. I remember just being mortified,' says the now 43-year-old. 'The next year, I wanted to get involved in the school play. I tried out but didn't get a part.' Then, when he was 16, his voice cracked while he was reading to the class. He was terrified he'd be bullied if it happened again. 'When we took turns going around and reading out of books, I'd look about three or four people ahead and go, "I need to go to the bathroom".'

Later, he went through a phase of sipping alcohol before speaking at university or work. 'I was never drunk; [I drank] just enough to silence the voice inside that kept saying, *You can't do this, you're going to embarrass yourself.*' That same inner critic stopped him being a best man twice and prompted him to turn down a director role at work.

In the United Kingdom, a survey in 2023 found public speaking was the second-most common fear, after heights.

'When our species was evolving, we would hang out in bands or groups of about 150 people, and your relative status in that group was absolutely critical,' says Abrahams. A higher status meant shelter, food and finding a mate; a lower status could mean death. 'Anything you did that risked your status, like getting up in front of people and doing something embarrassing or wrong, could have dire consequences.'

In the modern era, says Sarah Denholm, 'it's fear of harm, it's fear of threat, to our persona, to our ego, to certainty. It's that loss of control, it's vulnerability, it's all of those things that tap in so strongly to who we are.' Justin, the construction worker, doesn't think anyone notices how afraid he is, 'but inside, I'm freaking out. It's just being in that quiet setting, with eyeballs beaming on you; you think you're going to stuff up.'

> 'It's fear of harm, it's fear of threat, to our persona, to our ego, to certainty.'

Public speaking anxiety even has a name—glossophobia— from the Greek for 'tongue' and 'fear'. The trouble for many speakers is that a flight-or-fight response kicks in—sweating, nausea, dizziness, a dry mouth—just as they need to appear calm and in control. In severe cases, people 'choke' or even have a panic attack.

A shy and socially anxious business student, Kylie Campbell felt uncomfortable speaking with more than a couple of friends at once. During group presentations, the only words she ever volunteered to say were, 'Any questions?' Even then, she says, 'I was dying'. Terrified of the prospect of presenting solo, she sought help at public speaking club Rostrum, a non-profit founded in the 1920s in the United Kingdom. The first time she went to a meeting in Melbourne, she could speak in front of the group for just a few seconds. 'I didn't think anyone wanted to hear what I had to say.'

A surprising number of people in the public eye can come unstuck over public speaking. Roman statesman Cicero, famous for his orations, once froze during a speech at the Forum. 'I turn pale at the outset of a speech and quake in every limb and in all my soul,' he wrote. Actor Emma Watson had appeared in 12 films by 2014 when she told *Elle* magazine she was so nervous giving an address on gender equality at the United Nations headquarters that she wondered, 'Am I going to have lunch with these people, or am I going to be eaten?' Harrison Ford once told the *Los Angeles Times* his 'greatest fear' was public speaking.

'I turn pale at the outset of a speech and quake in every limb and in all my soul.'

'It's a sense that people are going to expect me to be this person,' says psychologist Corrie Ackland at the Sydney Phobia Clinic, 'so it's really going to surprise and shock them if they notice me to be anxious. That's going to be an affront to my reputation.' Sean McCaul was haunted by self-doubt before he eventually sought help with his public speaking. 'The hardest thing was feeling a bit like a fraud,' he says. 'I'm a very confident person in most situations.'

HOW DO YOU MANAGE YOUR FEAR?

Patrick Cripps, captain of the Carlton Blues in the AFL, says he was 'like a two-stroke motor trying to talk' in interviews during the 2015 and 2016 seasons. He remembers being lost for words on live television and radio. 'From this, I probably feared it more,' he tells us. 'But like any fear or phobia, if you try to push it away and ignore it, it just gets bigger.'

'Like any fear or phobia, if you try to push it away and ignore it, it just gets bigger.'

He stopped giving interviews for six months and consulted with a media manager, a senior player and a psychologist. Through mock interviews, he built up a repertoire of answers to various questions. 'My biggest piece of advice is that it will take time and practice,' he says. 'Once you get exposure to the situation again and you get through it, then the confidence starts to grow.'

A good way to manage fear is to be prepared. 'The paradox of practice is that it equals freedom in the moment,' Denholm says. 'If you know [your subject] extremely well, even if it's rehearsed extensively, so long as you are connected to that audience, you will not go into robotic mode.'

Think about the goal of your speech, suggests Stanford University's Abrahams. When preparing a talk, he asks himself: What do I want people to know when I'm done? What do I want them to feel? What do I want them to do? He says every speech needs to have a beginning, middle and end that connects ideas. 'One of the big problems with public speaking, especially for nervous people, is they end up just listing information. They take us on a journey of their discovery of what they're thinking about saying as they say it instead of packaging it up nicely.'

A best man speech, for example, can follow a simple structure: how you met the groom, how they met their partner, some words of wisdom, well wishes and a toast. The goal is to offer insights about the groom and to celebrate the couple (not yourself). Everyday stories are gold: the time the couple thought they were going wine tasting but the groom accidentally booked bocce lessons instead, and so on. Sprinkle in some humour but keep it tasteful.

'There is a lot of conflicting advice about public speaking,' Irvine notes. 'On the one hand, we are told to practise, practise, practise, and that practice makes perfect; but on the other hand, we are told to speak in a more natural, conversational manner, not to read or memorise.' It's all a matter of context, she says. A tight structure with linking statements, so the audience can follow your thread, helps in a formal setting. But 'speaking up at a staff meeting, talking through an idea in a team meeting or even pitching an idea to clients or colleagues, we want to be planned but not committed to an exact work order'.

In any case, most people can't memorise a full speech. For her lectures, Irvine prepares manageable chunks of information so she can rearrange or expand on points depending on how she thinks students are responding.

With your material sorted, you can 'rehearse' in front of a mirror, record yourself and even create distractions. Says Denholm: 'The hardest thing is to put on the TV or the radio—there's got to be talking, not music—and continue to talk while the volume is loud enough.' Doing push-ups before you speak allows you to practise with an increased heart rate. At his clinic, Ackland uses virtual-reality headsets to help people visualise themselves in an auditorium or at a wedding reception.

In the end, avoid perfectionism. 'I've had people get up and give a presentation. I think it's good, everybody in the room thinks it's good,' says psychologist and speech trainer Catherine Madigan, 'and all of a sudden, [the speaker] will throw up their hands and say, "Oh, it's terrible!".'

Olena Staroshchuk can relate to this attitude. After moving to Melbourne from Ukraine in 2014, she opened a women's clothing store but was very hard on herself because she struggled with customer banter in English. 'You can't be a salesperson and not be able to talk to people,' she says. 'I was this high-achieving girl at school and, for me, making a mistake was a disaster.'

Public speaking helped her to stop worrying about how people perceived her accent or grammatical missteps. Giving it her best, she found, was better than failing. 'I might not do it 100 per cent but I can learn something from it. Making a mistake is the best learning.' She is now the president of Rostrum Victoria.

Similarly, becoming too fixated on goals can be counter-productive. Abrahams' students 'want a good grade', and the entrepreneurs he coaches can be nervous about winning funding. 'If you can bring yourself to be more present-oriented, not be worried about the future, then you're going to be less nervous,' he says. Quick exercises before speaking such as counting backwards or saying tongue twisters are a way of focusing on the present.

And try not to take yourself too seriously. The clichéd advice for novice speakers is imagining their audience naked or in their underwear—'I can't imagine anything more nerve-wracking,' Abrahams says—but other visualisations can help keep things light. Denholm was coaching a chief financial officer when he said that he would imagine himself as a gorilla when speaking. 'I said, "Okay, you're a sparkling gorilla" and he loved it,' she says. 'He imagines himself as a sparkling gorilla and he laughs at himself.'

HOW DO YOU GET THROUGH THE ACTUAL SPEECH?

Abrahams became interested in public speaking anxiety after reflecting on a faux pas in his youth. At a public-speaking competition he decided to talk about his favourite topic: martial arts. He wanted to kick off the talk, literally, with a karate manoeuvre. 'I was so nervous I forgot to put on my special karate pants. I ripped my pants from zipper to belt loop in the first 10 seconds of a 10-minute presentation,' he recalls.

'Funnily enough, I got through the speech. The woman who was chaperoning the event threw me her sweater and, as I'm speaking, I tied it around my waist. I think because the parents [judging the competition] took pity on me, I actually won it.' In the moment, he'd flipped a mishap into comedy. 'We can do a lot mentally to reframe the way we approach our speaking to actually make us feel more comfortable and confident,' he observes.

Anxiety is often expressed through body language: crossed arms, clenched fists, inability to make eye contact. Abrahams coached a senior executive who, as he spoke, peeled the labels off water bottles and unscrewed pens until the springs flew out. 'He built a wall around himself with erasers,' Abrahams says. 'It was mostly his nervous energy.'

That kind of energy can be redirected. 'If I'm standing on a big stage presenting in front of a large audience, I can take a step forward with big broad gestures and welcome everybody,' Abrahams says. 'It gives my body a place to go; I'm moving, which is what adrenaline is trying to make me do, but I'm doing so in a way that people perceive as competent and appropriate versus the person who retreats and paces back and forth.'

Channelling nerves into enthusiasm won't work for everyone, notes Irvine. 'Instead, I prefer to think, *This is the first time my audience has heard my message.* This helps me to start at a more meaningful pace. I then rely on my structure and a short, sharp opening so I don't trip over my words and can ease into the speaking opportunity.' Some speakers start by asking the audience a question or playing a video. 'This takes the initial focus off the speaker and, again, allows them to settle into the talk or presentation,' Irvine says.

Asking a question is also a handy trick if you lose your train of thought, says Abrahams. 'Just knowing you can do that reduces the pressure because you know you have a ripcord you can pull that gets you out of that situation.' And if you rely too much on 'ums' and 'ahs', try a short pause instead.

Humour is 'the fastest reframe for the brain', adds Denholm. Jokes can work in unlikely situations: sensitively placed among poignant anecdotes in a eulogy, say, or to offer a breather before presenting charts and figures at an annual meeting. The safest humour is self-deprecating. 'I just take the piss out of myself,' says footballer Patrick Cripps. When he accepted the Brownlow Medal in 2022, he poked fun at himself for being unfit when he joined Carlton. 'A lot of people said I rolled into the club,' he said to laughter. 'I wasn't the most athletic bloke going around.'

And a word to the people listening to a speech: it is, to a degree, a group effort. Rather than staring blankly at the speaker, it helps them if you appear engaged. 'For anyone who's slightly shaky, no feedback is terrifying,' Denholm says. Put away mobile phones, don't whisper to your neighbour, instead smile, nod and laugh at the speaker's jokes. 'It's just doing those little conversational cues as if it was a one-on-one.'

WHY BOTHER TO SPEAK IN PUBLIC?

Singer-songwriter Meg Washington developed a stutter when she was four. By her final year of school, she couldn't deliver an English oral presentation. 'It was excruciating,' she tells us. 'It's like a bit of a tic. It comes out when I'm more vulnerable.' Over time, she learned to replace words she couldn't say: 'You might say the road instead of the highway.' This works, Washington says, until she gets to a proper noun such as someone's name. The only way she could completely stop her stutter was to have her voice follow a rhythm, such as when she sang. 'It was magic to me.'

'I have spent my life up to this point ... living in mortal dread of public speaking.' Washington won two ARIA Awards in her twenties, but public speaking remained her worst nightmare. Then she was invited to deliver a TED Talk about her stutter to more than 2000 people at the Sydney Opera House. 'The truth is I have spent my life up to this point, and including this point, living in mortal dread of public speaking,' she says during the speech in 2014. She stutters dozens of times and finishes to thundering applause.

The speech has been viewed online more than two million times. 'I certainly have felt totally liberated by that speech,' Washington says. 'The worst thing that could have happened did happen, which is that I stuttered.' Nowadays, she doesn't always attempt to hide it. 'If the stutter is there, then it's there.' She's presented awards, voiced a character on the children's series *Bluey* and narrated a TV show. 'I have just done all these things with speech I would have never been able to do.'

Washington's experiences bring to mind King George VI, who famously struggled with a stammer and eventually started seeing Australian speech therapist Lionel Logue. 'My dear Logue,' a delighted George wrote days after his 1937 coronation, 'The Queen and I have just viewed the film of our Coronation ... You know how anxious I was to get my responses right in the Abbey ... but my mind was finally set at ease tonight. Not a moment's hesitation or mistake!' Logue, played by Geoffrey Rush in the film *The King's Speech*, coached the then-Duke of York through breathing exercises, reciting tongue twisters and shouting vowels from an open window. In the film, he 'conducts' the King through a near faultless broadcast to the nation after war was declared on Germany in 1939. Afterwards, Logue tells the King: 'You still stammered on the w.' George, played by Colin Firth, replies: 'Had to throw in a few so they knew it was me.'

Sean McCaul has practised speaking in front of small groups at Rostrum club meetings for seven years. 'It's given me the opportunity to do speeches without the pressure I felt at work,' he says. His biggest regret is that he didn't do it sooner. 'I let it limit what I could've achieved in my career.' McCaul still gets nervous talking in front of people but enjoys pushing through. He even appeared on a panel speaking about his business. 'Considering how stressful it is, you do get a confidence boost after you do it.'

Many speakers describe experiencing a buzz after delivering a successful speech. Denholm explains this as feeling your energy land and then come back at you through the audience's response. 'And our system knows that. Our cells are vibrating, and we have this sense of resonance when it's working.'

Kylie Campbell once struggled with shyness but can now speak in front of packed auditoriums. When she feels a speech has worked, she walks away with a tingling sensation. 'I feel like I've actually done it,' she says. 'Seeing that people do listen and want to learn from what I've got to say, it's a good feeling. You're not just getting up there to fill some space and time, you've gotten a message across.'

With his friend's wedding on the horizon, Justin was slowly building confidence at a weekly public-speaking class. 'The best thing I've done is just to do it. I would avoid it any way I could in the past, and now I'm obviously doing this course and getting up in front of people,' he says. 'The more you do it, that's the only way you're going to overcome it.' He's looking forward to raising a glass to the bride and groom, his words lingering in the room for a moment, followed by applause.

7

WHAT'S DIWALI?

More than a billion people celebrate Diwali. What does it involve and who can join in?

Jewel Topsfield and Rachael Dexter

When Rujuta Limaye moved with her family to Australia in the 1990s, Diwali was mostly observed in the homes of the Indian diaspora. But now the five-day event, one of the largest religious festivals in the world, is celebrated in schools, offices, parks and stadiums across the nation. 'My mum works for a corporate and they're having a "Dress up and bring your own Diwali food" day and a Diwali-themed quiz,' says Limaye, who was born in Maharashtra in western India.

Around the world, more than a billion people observe Diwali, or Deepavali. It is as important to Hindus as Christmas is to Christians. In Australia, more than a million celebrate it, along with the Nepalese version called Tihar and the Sikh festival Bandi Chhor Diwas with which it coincides. 'Diwali celebrations are a lot bigger here now,' says Limaye. 'There's just more awareness of other places and cultures. I think Australia is a lot more welcoming than it was 30 years ago.'

'It's about good winning over evil or light over darkness.'

The meaning behind Diwali is beautiful, she says. 'It's about good winning over evil or light over darkness. I think that's universal, no matter where you come from.'

What are Diwali's origins, rules and traditions? And can you celebrate it even if you're not religious?

WHO OBSERVES DIWALI?

Diwali, also known as the festival of lights, is India's most important holiday. Mentioned in multiple ancient texts such as the *Padma Purana* and the *Skanda Purana*, it's thought to be a fusion of Indian harvest festivals dating back more than 2500 years. Today it has different meanings for different religious communities. 'Although Diwali originates from Hindu traditions, its significance resonates far beyond its religious roots because it is not confined to only one faith,' says Dr Surjeet Dhanji, a former director of cultural diplomacy at the Australia India Institute. 'It's a celebration shared by Hindus, Sikhs, Jains and Buddhists alike.'

For Hindus, more than 680,000 of whom live in Australia, Diwali marks the triumph of deity Lord Rama over the evil spirit Ravana. At this time, Sikhs—there are more than 210,000 in Australia—commemorate the release from jail in about 1619 of Guru Hargobind Singh and other prisoners incarcerated by a Mughal emperor. The Jains celebrate the enlightenment of Lord Mahavira, the founder of Jainism. Newar Buddhists commemorate Emperor Ashoka's conversion to Buddhism in the third century BC.

'So this is what truly makes Diwali remarkable,' says Dhanji, who is a Sikh. 'It's the ability to unite diverse communities coming together to light up their own and their communities' lives with lamps and candles, embracing the festival spirit of hope, looking beyond our differences and celebrating our shared humanity. In effect, it's really reinforcing the value of love and togetherness.'

WHEN AND HOW IS IT CELEBRATED?

The exact dates of Diwali vary each year depending on the Moon's cycle, but it usually falls in October or November. 'It's a very auspicious month in the Hindu calendar; it is called the month of Kartik,' says Reverend Albert Lange, a former chairperson of the Faith Communities Council of Victoria, who is an adherent of Vaishnava Hinduism.

The rituals and festivities differ in various places, but generally go for five days. Day one is known as Dhanteras, says the Hindu Council of Australia, the day of fortune that marks the arrival of the goddess of prosperity, Lakshmi. In Hindu mythology, Lakshmi is said to emerge from an ocean of milk when it is churned by the gods and demons.

Next is Naraka Chaturdasi, the day of knowledge, when the demonic Narakasur was slain by Lord Krishna. The third day is Deepavali, the day of light, when Lord Rama returned to the Indian city of Ayodhya after vanquishing Ravana, a demon king. Day four is Govardhan Puja, which the Hindu Council of Australia says marks 'Lord Krishna's feat of lifting the Govardhan mountain on his little finger [like an umbrella] in order to save the residents of Vrindhavan from the torrential rains'. Finally, Bhai Duj is the day of love between siblings, when brothers and sisters pray together.

The word Diwali comes from the Sanskrit *deepawali*, meaning 'row of lamps'.

The word Diwali comes from the Sanskrit *deepawali*, meaning 'row of lamps'. In the lead-up to the festival, many people clean and decorate their homes, temples and offices. They also light oil lamps made of clay called *diyas* to welcome Lakshmi into the home. 'The oil lamps emanate a lot of positive energy,' says Srinivas Shesham, who has organised Diwali festivities in a Melbourne park since 2013. 'This period is very spiritual and emotionally healing for all people involved.' Many families give sweets called *mithai* and buy clothes and presents—in India the festive season provides a boost to the economy.

The weekend before Diwali in 2023, Limaye, her mother and her aunt prepared a feast that included a deep-fried savoury snack called *chakli* and sweet coconut-filled dumplings called *karanji*. The family also drew *rangoli* patterns—intricate designs using coloured powders, rice or flower petals—in their front and back yards. 'Each day you can make a different *rangoli* pattern,' says Limaye.

On the main day of Diwali, families gather to pray, eat, sing, dance and share gifts. 'Like Christmas, there's a religious and spiritual aspect,' says Albert Lange. 'It's a camaraderie and love for one another. That's the kind of thing that appeals to most people. And also the remembrance of those past times of the Supreme Lord. So it brings that spirituality.'

Diwali is also often marked with fireworks. 'If you go to India, it's like World War III—there're so many different fireworks and crackers and rockets,' Lange says. The smoke adds to India's already toxic air, so New Delhi authorities have banned firecrackers in recent years.

WHERE IS DIWALI CELEBRATED?

All over the world. As well as in India, the main day of the festival—the day of Lakshmi Puja—is an official holiday in many countries including Fiji, Malaysia, Mauritius, Myanmar, Nepal, Pakistan, Singapore, Sri Lanka and Suriname in South America. It was a public holiday in New York City for the first time in 2024, when all public-school students had the day off.

The White House first marked Diwali in 2003 under then US President George W. Bush and celebrations were scaled up after Kamala Harris, who is Black and Indian American, became vice president in 2021. Former British prime minister Rishi Sunak, who is a Hindu, was elected to the top job during Diwali in 2022. Two years earlier, as chancellor, he had marked Diwali by lighting candles on the doorstep of No. 11 Downing Street, telling *The Times* it had been one of his proudest moments.

Surjeet Dhanji says the Indian diaspora would like Australia to declare a public holiday during Diwali. 'So it should be for Eid [which marks the end of Ramadan] as well,' she says. 'Those who observe religious faith holidays should be allowed to take a day off.'

CAN ANYONE GO TO DIWALI CELEBRATIONS?

Hundreds of Diwali events are held across Australia each year, from small local temples to major stadiums. The public festivities are open to everyone, says Shesham, who organises Diwali celebrations in Melbourne's west. He remembers being delighted that the first family to arrive at the inaugural Diwali celebration at a park in suburban Wyndham Vale in 2013 was European. 'I thought, *okay, this is a very good sign that the universal theme has really registered*,' Shesham says. 'Since then, I would say about 40 to 50 per cent of the families who come to the event are from other communities.'

'On Diwali we open up our doors to all faiths.'

Lange says many of the Indian diaspora invite friends to celebrations. 'On Diwali we open up our doors to all faiths and our Muslim brothers will come, our Christian brothers will come to celebrate Diwali. And, similarly, on Ramadan or Easter or Christmas, the reverse would happen. That brings about social cohesion in difficult times . . . We need that companionship of each other, whatever tradition we're in, and we put our political beliefs aside, and just come together in friendship and love.'

Many of the most popular events are held in communities with large numbers of Hindus, such as Blacktown in Sydney's west, where the council has run a Diwali lights competition for residents since 2017. Lange says people attending events hosted by councils or community groups will be 'blown away' by the sweets. 'They're absolutely delicious. You'll be confronted with a lot of food, friendly people dressed up in their finery, lots of singing and Bollywood dancing.'

Or people enjoy Diwali at home. 'The traditions are quite varied across the country,' says Limaye. 'I feel there's a freedom to celebrate it my way and what suits my family in Australia. It's important because it's a sense of where you come from and passing that on to the kids.'

8

WHAT'S VERTIGO?

It can put you in a spin for hours, days or longer. If you thought the Hitchcock film was unsettling, try living with this real-life condition.

Samantha Selinger-Morris

The day that started Kathee De Lapp's downward spiral from being a well person to one who needed to learn to walk again began like any other. She drove to her morning university lectures, spent eight hours in classes and drove herself home. That afternoon, she felt she might be coming down with a cold. 'You know, I felt just kind of lethargic, just generally yuck,' she says.

The next morning, when she rolled over in bed to get up, the room began to spin. 'I could hardly stand straight or walk straight,' recalls De Lapp, who was then 29. '[I was] vomiting for 24 hours. It was very drastic—black and white—from being normal to being completely incapable.'

Despite persistent symptoms, the problem wasn't correctly diagnosed for seven years. She was suffering from vertigo. 'I couldn't walk across the room,' says De Lapp, now a neuro-physiotherapist who treats people with balance disorders. 'I essentially had to learn to walk again and navigate and do things. It is an unkind, life-altering problem.'

Up to one in three Australians will suffer vertigo. And while it strikes most people intermittently, ongoing symptoms such as De Lapp's are not unheard of. What causes vertigo? Are certain people susceptible to it? And can it be cured?

WHAT ARE THE SYMPTOMS OF VERTIGO?

In the 1958 movie *Vertigo,* directed by Alfred Hitchcock, a cop played by James Stewart is scared of heights. He quits his job after a rooftop chase sees him dangling, wide-eyed with terror, from a flimsy gutter high above the street. The movie features a lot of terrifying falls (falling from a tower, falling in love). In reality, though, vertigo is not the same as a fear of heights—that's acrophobia. Nor is vertigo a condition in and of itself. It's a symptom that can be caused by a variety of factors and which creates the illusion of movement when there isn't any.

'Some people give fairly interesting descriptions of walking along and it's as if the footpath is [moving] or on an angle, and they know it's not,' says Associate Professor David Szmulewicz, a neurologist and neuro-otologist (a physician who treats ears) at the Royal Victorian Eye and Ear Hospital in Melbourne. 'Others will describe a sensation of, "I turn my head, and it takes a few moments for my brain to catch up".'

Some people feel as if they're moving from side to side or up and down. Others say they feel as if they're on a rocking boat, the room is swaying or their head is spinning. 'There's, generally speaking, the idea that if you spin, it's vertigo, and if you're not spinning, it's not vertigo,' says Dr Luke Chen, a neurologist and neuro-otologist at Monash and Alfred Health in Melbourne. 'That's technically not correct.'

These feelings might not sound too terrible. But the symptoms can last for hours, days or even weeks, depending on their cause. And because dizziness can cause vomiting, vertigo often leads to persistent throwing up. (Nerves in the ear are directly connected to the parts of the brain that trigger nausea, explains Szmulewicz.) 'You have no idea that anything could feel so earth-shatteringly wrong,' says De Lapp, who treats patients at the Hearing and Balance Centre at St Vincent's Hospital in Sydney. 'But it does.'

The symptoms can last for hours, days or even weeks, depending on their cause.

WHAT CAUSES VERTIGO?

Vertigo is often caused by damage to the organs deep inside our ear that make up our vestibular system, which largely regulates balance. At the heart of the system are three small semicircular canals. As we move our head, fluid moves along the canals much like water in a bucket when you tip it. Nerves in the canals then send rapid-fire messages to the brain at the rate of 20 per second—in particular the cerebellum, which controls coordination—about how far, fast and in what direction the head is moving. 'So the ears are messengers,' says Chen. 'Now, the brain's going to have to use that information and say, "Okay, the head is moving in that direction, so how do we make our next step?".'

Our vestibular system operates a bit like a stereo, with our left and right ears sending separate signals to our brain. They need to be precisely coordinated for us to remain balanced. If one ear becomes damaged—say, by infection—the signals fall out of sync. Our eyes then move around to make sense of the incompatible information. This involuntary rapid eye movement, either horizontal, vertical or circular, is called nystagmus. 'Your eye goes wobble, wobble, wobble, wobble,' says De Lapp. 'You still see, there's no blindness, but all that movement makes no sense at all . . . And that looks like the room's spinning. So the room spins, and if it does that for five seconds it only takes . . . a fraction [of a second] for you to fall down.' Or, as David Szmulewicz, puts it: 'Your world's not moving . . . your eyes are moving your world around.'

About half of vertigo cases are known as benign paroxysmal positional vertigo, or BPPV. The faulty messages to the brain are caused by microscopic crystals in the inner ear that become loose. Normally, the crystals, which are made of calcium carbonite, are fixed in jelly in our ear canal. 'They're a bit like cement between bricks; they keep the lining of your inner ear together,' says Luke Chen.

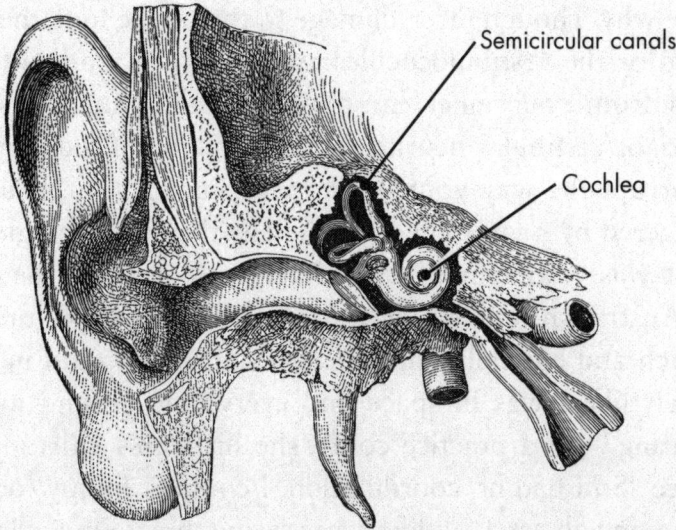

Semicircular canals

Cochlea

Fluid in the semicircular canals triggers messages to the brain.
Hein Nouwens/iStockphoto

Even a seemingly benign action can set them adrift: tilting your head suddenly while turning over in bed or leaning down to pick up laundry from a basket. The rogue crystals cause nerve endings to send the faulty signals, and the spinning begins.

Rogue crystals cause nerve endings to send the faulty signals, and the spinning begins.

In the case of Australian golfer Jason Day, it was: take a step, look down and collapse. That's what happened during the US Open in 2015, when he was struck down by an attack of BPPV. 'The vertigo is a difficult thing, it just comes and goes whenever it pleases,' Day told CNN. 'I've had it before and there have been years between stretches and, unfortunately, it happened at the US Open and that knocked me off my feet.'

Meniere's disease, which involves a build-up of fluid in the inner ear, is another cause of vertigo. Migraines are sometimes linked to vertigo, though doctors aren't entirely sure why. Though rarer, damage to the cerebellum, the brain stem or the vestibulocochlear nerve, which sends information from your inner ear to your brain, can also set off vertigo. Vestibular neuritis, where swelling of the nerve can interrupt the way your brain reads information, is usually triggered by a viral infection such as the flu or shingles.

It was vestibular neuritis that nearly derailed the career of Australian tennis player Alicia Molik. 'I woke up for a match and basically fell into the wall,' Molik said in 2005. 'I felt like I was in space and everything around me was floating.' On a practice court, she hit tennis balls into the fence. 'So I had no coordination. I couldn't fix my focus on a moving object. I felt like I was seeing things almost in third person. It was really scary.' It took her two years to 'fully come through' her vertigo. The vertigo, along with injuries, contributed to her decision to step back from the game in 2008. (She later made a comeback.)

De Lapp's vertigo was eventually traced to a stroke caused by a benign tumour in her brain. 'I had an internal bleed, and it bled into that part of the brain stem where the signal balance goes,' she says.

A rare cause of vertigo is a syndrome called superior canal dehiscence syndrome (SCDS). A hole in the bone that encases the ear's vestibular system creates a phenomenon called autophony, the ability to hear your bodily processes, from your heartbeat to your eyes clicking (yes, clicking) as they move from side to side. 'I often describe it as a sort of bionic hearing,' says Szmulewicz.

Sufferers may hear footsteps very loudly from a great distance away. In years gone by 'these people were told their symptoms were imagined,' says Chen. 'These symptoms were put down to a psychiatric problem.'

To make matter worse, the canals of the vestibular system may slip through the hole in the bone. 'Part of what happens is that you're converting the balance system to be able to respond to sound, which it normally doesn't,' says Szmulewicz. 'So people get noise-induced dizziness.'

The risk of vertigo increases as we grow older. Certain parts of the vestibular apparatus age, which may increase the chances of crystals being shed in the inner ear. And people with osteoporosis are believed to have a higher likelihood of loose inner ear crystals and therefore BPPV.

HOW DO DOCTORS KNOW YOU HAVE VERTIGO?

Often the key to an accurate diagnosis is a person's eyes. 'The eye movements are breathtaking, I have to say,' says Szmulewicz. A doctor might ask a patient to lie down and ask them to move their head to the side in a way that is likely to trigger vertigo. 'There's a whole set of eye movements that point us to [damage to] the cerebellum and its connections,' Szmulewicz says. For instance, if a person's eyes jerk to the left when they look left and jerk to the right when they look right, the cause is likely a lesion in the brain, usually the cerebellum. To diagnose BPPV, doctors look for jerking eye movements that move up and down, side to side, or in circles.

CT and MRI scans of the inner ear can detect other causes of vertigo, such as structural damage or inflammation. Some infections that cause vertigo, such as labyrinthitis, are also associated with hearing problems—so a hearing test might help.

'You definitely can treat most causes of vertigo very well, once you get the correct diagnosis,' says Chen. This is not always simple. When patients tell doctors the room is spinning, they might be tested for evidence of a stroke or even cancer. If they get the all clear, they might be given a Panadol or anti-nausea drugs. 'You exclude the scary causes and then you go no further,' says Szmulewicz. 'But you're not offering them a diagnosis and, in general, the best treatments are specific for diagnoses.'

Some vertigo sufferers report not being taken seriously. 'That's the bane of my day-to-day practice,' says Chen, whose patients are often at the end of their tether by the time they see him. In part, this is because they weren't experiencing a vertigo attack during earlier examinations. 'So there's the natural tendency to think the person would be a bit hysterical or embellishing their symptoms, if you like,' says Chen. 'They do get dismissed . . . and I suppose there's always a bit of stigma, that, well, you've just got to put up with it.'

'Vertigo is perhaps one of the hardest neurological symptoms to deal with.'

Patients can be misdiagnosed too. 'One of the common ones is they have migraine vertigo and [instead] they've been diagnosed with Meniere's disease,' says Chen. 'Vertigo is perhaps one of the hardest neurological symptoms to deal with. And there are many anecdotal quotes that even experienced [medical] practitioners will sigh at seeing someone with vertigo . . . They can feel daunted, or even intimidated, by someone who has complex vertigo.' (People can suffer from multiple causes of vertigo at the same time.)

Delays in a correct diagnosis can have knock-on effects. 'The studies have shown that 50 per cent of patients suffering from vertigo have anxiety and/or depression,' says Chen. 'That's clearly higher than the rate of anxiety and depression in the community [about 20 per cent].' He has even seen vertigo lead to relationship break-ups. 'Relationships with friends, family, work. They can't go to work. It's a very difficult condition to manage.'

HOW DO YOU STOP VERTIGO?

One option for people with BPPV involves a delicate physical treatment: a 'log roll' or an Epley manoeuvre (named after John Epley, the American specialist who developed it in 1980). A doctor, vestibular audiologist or physiotherapist moves the head in a particular way to coax the free-floating crystals out of the inner ear canals and into the vestibule where they no longer cause havoc. (A patient can be taught to do this at home.) The treatment can also be delivered with an Epley Omniax machine, which look a bit like an amusement park ride: a patient is seated inside and moved gently upside down and in circles. Infrared goggles record their eye movements. Computer software determines which canal is affected by the moving crystals, and the clinician resets the patient's inner ear by moving the machine in a particular way.

BPPV can be prevented by avoiding movements that trigger the vertigo. If a person's vertigo was triggered by a virus, as is the case with vestibular neuritis, medication can reduce dizziness, inflammation and nausea, as can vestibular physiotherapy. 'It is about teaching the brain to keep the gaze co-ordinated properly,' says De Lapp. With Meniere's disease–induced vertigo, drugs can help alleviate nausea, vomiting and the spinning feeling.

Amazingly enough, our brains also naturally compensate for imbalanced messages it receives from our ears.

'The brain senses that there's a difference between the two ears, and that's the start of a process we call compensation,' says Chen. 'That refers to a series of chemical changes at different levels of the brain which are aimed at restoring, if you like, the balance between the two ears . . . so with time, a person's balance improves.'

Longer term, the cure depends on the cause. Dislodged crystals can never become fixed in their jelly again. Some BPPV patients 'have periods where they're absolutely fine', says De Lapp. 'But then they go to pick up something off the floor, and they might fall over.' Meniere's disease, too, has no cure. Fortunately, SCDS (where you can hear your heartbeat) does. Surgeons simply patch the hole in the inner ear bone. And Kathee De Lapp hasn't had vertigo since a doctor removed her tumour in 1991, seven years after she suffered her first symptoms.

9

WHY DO PEOPLE QUEUE FOR BRUNCH?

You'll see them on weekends, the long lines
outside feted eateries. Some people
even look like they're having fun.
What are they thinking?

Osman Faruqi, Felicity Lewis and Carla Jaeger

It was the spring of 2020, and a new cafe had flung open its doors in suburban Sydney. 'We thought it would take some time to build a bit of a reputation,' recalls co-owner Chris Theodosi, 'and we would have loved, after a year or two of, you know, selling good food and good coffee and happy vibes, to have become a bit of a destination. But it happened a lot quicker than that.' Within minutes on that first day, the cafe was packed and Theodosi was jotting names and phone numbers on a waitlist.

Today, Happyfield is a lively fixture on a busy corner in inner-west Haberfield. 'Pretty much every Saturday and Sunday since we opened, we go through a fair few pages for that waitlist,' says Theodosi. 'I think, generally, a busy cafe, or a cafe with a waitlist or a queue, definitely works in your favour 95 per cent of the time. I think [for] most people, if they're a bit of a foodie or just looking for a good thing, it definitely attracts more people.' To avoid clogging the footpath, would-be patrons are encouraged to go for a walk and wait for a call from Theodosi. 'If every single person on our waitlist was waiting out the front, it would be chaos,' he says. 'It's still chaos. But it's organised chaos.'

Someone once observed that the British love a queue, but clearly so do some Australians—just visit Happyfield, or The Grounds of Alexandria in Sydney, or Melbourne's Lune Croissanterie on a Saturday morning. Queuing culture, particularly outside buzzy eateries, has well and truly taken off. Lining up on the weekend for a $50 brunch that might consist of a couple of poached eggs and a flat white used to be limited to the trendy inner city, but it's now a common occurrence across the suburbs. Good luck trying to get a roll from Harvey's Hot Sandwiches in Parramatta or a Korean 'hangover soup' from Yeodongsik in Lidcombe without factoring in a solid wait. Things have gotten so out of hand that food blogs in publications like *Time Out* now unironically publish lists of 'Food queues that are worth the wait'.

So, what's going on? Do we just love the order and process that queues bring? Or are we seeking out long lines to show off how willing we are to sacrifice our time for the perfect pastry, coffee or sandwich? Or is a queue an event in itself?

HOW LONG HAVE WE BEEN QUEUING?

The *Book of the Dead* in ancient Egypt described recently departed souls queuing in a hall of judgement. They would be ushered into eternal life or devoured by a creature that was part crocodile, part lion and part hippopotamus. So it was a nervous wait. Luckily, goddesses attended to them in the line, including the kindly Qebhet, who personified the cool water of the Nile. It was yet another example of the Egyptians being ahead of their time: offering a kind of pre-service to queuing customers is today a tried-and-true business ploy to soothe, distract and add value as the clock ticks on for what can indeed seem like an eternity.

'Queueing is just a way of rationing a scarce resource,' says cultural historian Joe Moran, professor of English at Liverpool John Moores University. Queuing hit its stride in modern times during the Industrial Revolution, when large numbers of people moved to urban areas, necessitating more structure around the distribution of goods. Lines of people became a common sight during economic crises, including the Great Depression, particularly for the urban poor, who relied on charity to help put food on the table.

The notion that queuing is synonymous with the British seems to have formed during World War II when rationing became a part of life for large swathes of the population, writes Moran in his book *Queuing for Beginners: The Story of Daily Life from Breakfast to Bedtime*. Of course, queuing for rations wasn't uniquely British. 'There's a particular mythology that developed in the Second World War,' Moran tells us. 'I think there's a slightly masochistic aspect to the mythology. There's, "Oh yes, we're very stoical. We put up with terrible boredom and tedium, and we're still very polite".'

In fact, many Brits did grit their teeth during the war, including the elderly and pregnant women, whose ration books were labelled *Queue priority, please*. 'Queuing then was very fraught because it took up so much time,' says Moran, 'and also because people were queuing for things they actually needed. They would often have to have police disperse queues that were getting rowdy.'

Winston Churchill was between stints as prime minister in 1950 when he warned that Britain would become a 'Queuetopia' if Labour was re-elected. 'If they have the power, this part of their dream will certainly come true,' he growled. Behind his warning lurked the spectre of Soviet breadlines. 'People stood in line for everything, for bread, sugar, nails, news of an arrested husband, tickets to *Swan Lake*, furniture, Komsomol vacation tours,' wrote novelist Vladimir Sorokin of post-revolution Russia in his essay 'Farewell to the Queue'. Queues became an 'important therapeutic ritual . . . honed and polished over the course of decades.' Alexei Sundukov's 1986 painting *Queue* shows fur-hatted women in a line that vanishes into the distance. Mikhail Gorbachev reportedly shared a queue joke with then British Prime Minister John Major in the 1990s: A man stuck in a queue says, 'I'm fed up with this. I blame Gorbachev. I'm going to kill him!' He storms off but returns later and pushes his way back in. 'What happened?' someone asks. 'The queue to kill Gorbachev was just too long.'

Queues took on a different meaning in Russia during elections in March 2024, when anti-Kremlin protesters lined up outside polling stations en masse in an action called Noon Against Putin, a plan endorsed by the opposition leader Alexei Navalny before his death a month earlier in an Arctic jail.

But the queues we're talking about, the jovial ones outside hip cafes and restaurants, aren't about rations, protest or turmoil. So, what's driving that urge to wait in line?

WHY DO WE QUEUE TODAY?

Queueing 'is my definition of hell', says a customer called James, as he waits outside a destination eatery in Melbourne. It's a clear Saturday morning, and the queue stretches down the street and around the corner. James tells us he has bad memories of queues from a visit to Disneyland. Today, though, he has made an exception to his 'no queue' rule as he's with friends and family, on holiday from Sydney, and they'd heard about the food at this place.

As it happens, Disneyland and Disney World were pioneers in managing queues (although James clearly didn't appreciate their efforts). The theme parks introduced now-ubiquitous 'switchback' (or zigzag) queues in the 1960s. They were designed to make lines seem shorter and save space while encouraging people to chat. Today, the parks' phone apps estimate wait times for rides (more on that later) and offer virtual bookings. The parks also made an art of turning queues into experiences. When Disney Resorts revamped its spooky Haunted Mansion ride in 2024, for example, it trumpeted 'an expanded outdoor queue to immerse guests in enhanced theming'.

In fact, a lot of thought has gone into queues over the years. 'Queueing theory' is the mathematical study of waiting in lines. Queues offer insights for sociologists studying 'social systems'. For businesses, there is 'queue furniture' (such as plastic bollards with retractable belts) to help maintain 'queue discipline'.

Still, there's a surprising lack of research into queues, say psychologists Adrian Furnham, Luke Treglown and George Horne. 'For many organisations, customer disgruntlement at waiting times is a serious issue that demands solutions,' they write in the journal *Psychology*. 'It leads to "queue rage", which is physical and verbal abuse as a function of even minor delays.' The trio reviews all sorts of experiments on queuing's social norms and optimal conditions. Studies have shown, for example, that telling customers roughly how long a wait will be makes them more accepting of the delay—they feel confident there's a system in place, even if the estimate is wrong. Does music make a wait better? Familiar tunes with a fast tempo, such as pop, seem to work best. What about scent? Lavender has been shown to calm frazzled customers in banks and medical waiting rooms. (Cafes presumably have an advantage here: aromas such as ground coffee or pan-fried chorizo tend to trigger reward circuits in waiting customers' brains.) How about a cup of tea? The owner of Yeodongsik in Sydney has the right idea: he hands out barley tea to queuing customers.

But why do we get into queues in the first place? There are a number of psychological reasons, says Dr Meg Elkins, a behaviour economist at RMIT University in Melbourne, not least that lines suggest that people are waiting for something desirable. 'Why else would people forgo their time to wait in a queue?' Queues have a heuristic function. 'Heuristics are mental shortcuts that we use to make decisions because we have to make so many decisions in our day. Let's say we're going past a restaurant. If we see a queue, it decides for us that it must be a good place to eat.' As Happyfield's Chris Theodosi notes, 'If you see a long line for a Vietnamese roll, you go, "Oh, I want to line up for that one." Even if there's three or four others in the same street, you want the one everyone else is lining up for.'

Moran agrees. 'Given that we are mimetic [mimicking] beings, we tend to follow what other people are doing. Yes, it might look a bit tedious to stand in that queue, but it's also advertising something that you can see lots of people want. In the Second World War, people would join a queue when they didn't know what was at the end of it because they reasoned, probably rightly, that there would be something they would want.'

It's advertising something that you can see lots of people want.

Even in peacetime, when most goods are plentiful, we wrestle with concepts such as FOMO (fear of missing out). Lune Croissanterie in inner Melbourne is known to sell out every weekend, which creates an impulse for customers to line up early to avoid going home empty-handed. It's an important persuasion tactic, says Elkins. 'We see a queue and think, *Well, there's a scarcity for that product*, so it triggers our desire to be persuaded to stand in line.' Some businesses deliberately structure their operations to encourage queues, such as minimising seating space to create a sense of exclusivity or limiting supply to generate a sense of urgency. Theodosi is the opposite: 'We're like a NASCAR crew when a table leaves.' Add the Instagram factor, where people post photos of themselves queueing at the latest trendy venue, and you've got a self-perpetuating hype cycle that fuels FOMO and just keeps the queue going.

And don't forget, we're talking about brunch, a melding of breakfast and lunch originally concocted as a breezy *social* event. 'Brunch is cheerful, sociable and inciting,' an article in British *Hunter's Weekly* said in 1895. 'It is talk-compelling. It puts you in a good temper, it makes you satisfied with yourself and your fellow beings, it sweeps away the worries and cobwebs of the week.' In a similar spirit, the *Queensland Figaro and Punch* asked readers in 1927, 'Why Not Brunch?', noting that the meal had been embraced by 'youth' who 'discovered some time ago that lunch, or the Victorian dinner, served between one and two, cuts up the day indefensibly'.

Customers John and Betina are waiting outside Lune, with the sweet smell of baking pastry in the air. 'I've never queued before,' says John, 'I've got other things to do.' He's never even lined up at a nightclub, he tells us. Once, he and Betina just walked away from a long queue for tickets to a musical festival and had a great time at a cocktail bar instead. This weekend, though, the Sydney couple is in Melbourne to visit their daughter Luisa. 'They talked me into it,' says John of queuing for croissants.

Sarah, David and their puppy Baxter drove half an hour to the croissanterie and are now standing outside it in the sunshine. When Sarah saw the queue, she'd thought, *Well, that's just ridiculous. Is it TikTok famous or something?* But friends recommended the place, and they want to try something different. Baxter seems sanguine.

Perhaps something else is at play. Despite the opportunity cost it entails—we could be earning income, watching a movie, playing sports—queuing shows that we have the time, the social and economic capacity to spare to gain a highly desired product. We might all want the most delicious croissant, samosa or gourmet bacon and egg roll, but only some of us have the ability to acquire them. In fact, even when the desired product is not particularly extravagant or expensive—a *banh mi,* or a meatball sandwich, or a hamburger— the act of queueing for it bestows some kind of intangible value, similar to what the philosopher and economist Karl Marx referred to as commodity fetishism. According to Marx, goods become fetishised when their value is unlinked from their purpose or the amount of labour that went into producing them.

Queuing shows that we have the time, the social and economic capacity to spare to gain a highly desired product.

Then there's the collective aspect of queues. At their best, queues can represent equity and transparency. 'One of the reasons people don't mind queues,' says Moran, 'is that they are a very fair and clear way of rationing your "scarce resource". So, if you are prepared to get up really early for your brunch, you'll be at the head of the queue.' Related to this is camaraderie, a 'we're all in this together' vibe. 'Queuing for a rock concert, queueing for brunch, queuing for Wimbledon—there's sort of a recreational aspect to it,' says Moran. Even when time drags on, Elkins notes, there's a community element 'because you're talking about the pain of the wait . . . and you've got this collective purpose'.

Having said that, it's not all unicorns and rainbows when a queue is the only thing standing between a peckish diner and their avocado with dukkah. 'Not everyone wants to go for a stroll while they wait,' says Happyfield's Chris Theodosi. 'They want to kind of keep an eye on things and make sure they don't lose their spot. But we run a pretty fair system. We've been offered bribes and all sorts of things and never taken one.'

But there are limits. No one will stand in a queue indefinitely. 'There's a tension between the queue demonstrating that something's worth queuing for and this pain point, which is, "Oh, yeah, I won't bother when it's that long",' says Moran. Back outside Lune, we ask patrons about their wait limit. 'I've already exceeded it,' grins one customer, who has only just joined the end of the line. 'Fifteen minutes—I'm too impatient,' says another. Others say they are willing to stand for an hour. (Studies have shown that, in deciding whether to walk away from a queue, people will count how many bodies are ahead of them, which is what will actually determine their wait time, yet will also be more inclined to stay if the queue stretches away behind them.) A group of younger people waiting outside Lune recount how they once got up at 5 a.m. to queue for a fancy new menu item at a cafe. Novelty is a factor, they agree, as well as finding out for themselves what the social media and word-of-mouth hype is all about. Queuing may even be seen as a kind of adventure. 'There's a social factor,' says Darren, who is catching up with friends Patricia and Evan. On any return visit, though, 'there's a certain threshold: half an hour'.

WHAT'S THE ETIQUETTE OF QUEUING?

'It's not fair that people are seated first come, first served,' says Elaine in an episode of *Seinfeld* as she waits with George and Jerry for a table at a Chinese restaurant in New York. 'It should be based on who's hungriest.'

A version of Elaine's suggestion was put to the test by a group of social psychologists in the United States in the 1980s. To see how members of a queue would defend its integrity, researchers cut into 129 lines, saying only, 'Excuse me, I'd like to get in there.' Mostly, they succeeded even if, in about half of the cases, people gave them the passive-aggressive treatment: glares, eye rolls, tut-tuts. They were physically blocked in just 10 per cent of cases.

But when there were two interlopers, 91 per cent of people objected. (Who hasn't let out a highly audible sigh when the person ahead of them turns out to have been holding a spot for a group of 12?) Still, the researchers concluded that a single-file queue was a relatively weak social order—people focused on their own small patch, and there was a built-in disincentive to defend it, given starting an argument might mean losing your place.

What if you ask nicely? In 1978, researchers cut into a queue at a photocopying machine (remember those?). When they said they needed to copy five pages, they were let in 60 per cent of the time. However, when the person butting in said they had to make 20 copies, people already in the queue were less obliging. Only 42 per cent said yes to this request: 'Excuse me, I have 20 copies to make. May I use the Xerox machine because I am in a rush?' So, think twice before trying to jump the queue on a Sunday morning with, 'Excuse me, may my five friends and I get in here because we're a little bit time poor?'

Etiquette guide Debrett's of London describes queue-barging as 'a serious offence', unsurprisingly, but also rails against the failure to keep things moving. 'Anyone who isn't fully committed to moving forward an inch for every inch that opens up will earn almost as much disapproval . . . as the shameless barger,' it declares. Moran notes: 'People have been complaining about the decline of queue etiquette for as long as they have been mythologising how wonderful the British are at queuing.'

'People have been complaining about the decline of queue etiquette for as long as they have been . . . queuing.'

On the other side of the ledger, it doesn't take a scientific study to show that cafe staff will quickly attract glares if they are seen joking around or otherwise not doing their best to keep a queue shuffling along (although there are studies that *have* shown this). Assuming everyone is on their game, a general rule of thumb is to follow the rules of the venue. Remember, this is a social contract you've voluntarily entered into. Do they want you to line up? Are you supposed to put down your name and number? Do they want you to leave and come back when they call you?

The most important thing is respect. Staff are doing their jobs and other customers are in the same situation as you are. By all means grumble—it can be a bonding activity— but don't be obnoxious. It's frustrating but accepting the terms of the queue is essential. And try to enjoy yourself. Do a quiz. Get to know the person next to you. Or, if it all gets too much and you really are very hungry, just walk away.

10

WHAT'S IT LIKE TO HAVE TOURETTE'S?

Most people with this disorder don't swear.
But their tics can be tricky to tame.
How do they do it?

Jackson Graham

Most of the time it's like an itch. The urge increases until you give in to it. Holding it back can feel like being told not to press a big red button; the need to press the big red button only intensifies. This is how Tourette's feels for James Sayers. He is sitting in his car explaining this to me. Then, mid-sentence, he sounds the horn, twitches in his seat and swears twice. 'I just imagined my horn like the button,' he says. 'And sure enough, I'm pressing it.'

His involuntary actions, called tics, happen in the blink of an eye. Other people with Tourette's (also known as Tourette syndrome) compare the onset of tics to a sneeze or a shiver. They can be routine or random, come and go with stress and fatigue, and involve anything from head nods or grimaces to grunts or words. Contrary to the disorder's portrayal in the media, swearing occurs in only a handful of cases.

With a conscious effort, Sayers, 33, can temporarily stop his tics. Eventually, though, they 'come out one way or another, like tenfold'. Tics usually develop during childhood and between a third and half of cases wane by adulthood. Related mental health and neurodevelopmental disorders, as well as the threat of bullying, can make school years particularly challenging. Year 12 was hardest for Sayers when he would involuntarily rip pages from textbooks. He was diagnosed with depression at the time.

People whose symptoms continue into adulthood often develop ways to live with the disorder and minimise its effects. So, what's it like to live with Tourette's? How do people with it lead successful lives? And how should you respond to someone with the disorder?

WHAT'S A TIC AND WHAT'S TOURETTE'S?

The Parisian noblewoman known as the Marquise de Dampierre would abruptly yell curses and insults while in church, at the theatre, out walking or visiting her aristocratic friends. Her mysterious case was noted by a physician in 1825 and discussed in medical papers and wider publications for decades. But it wasn't until 1885, shortly after her death, that her condition was named as one of nine examples of what neurologist Georges Gilles de la Tourette called *maladie des tics*. Tourette never met the Marquise, and his instructor, Jean-Martin Charcot, encountered her only in passing. (Her true name was Countess Picot de Dampierre.) Still, the case became 'a sort of shorthand description . . . of the disorder itself', notes Howard Kushner in his book *A Cursing Brain? The Histories of Tourette Syndrome*.

In fact, only about 10 per cent of people with Tourette's experience uncontrolled bursts of swearing (called coprolalia, from the Greek for 'faeces' and 'speech'). Sayers has been yelling 'bitch' for as long as he can remember. 'It's now part of my friendship group; if I call you bitch it means I love you because we've kind of turned it into something good.' Other obscene, and rare, tics include copropraxia (inappropriate actions or touching) and coprographia (inappropriate writing or drawing).

Far more common are motor tics: blinks and facial twitches, or jerks elsewhere in the body. Stressful situations cause Tim Usherwood, a retired GP with Tourette's, to noticeably twitch and shrug his shoulders. 'I still have a wonderful life and a fascinating and challenging career,' he says. 'I see it as a part of who I am.' He also experiences vocal tics such as grunting. Other people with Tourette's might sniff, cough, make high-pitched sounds or repeat words.

Tics can shift over time and echo what people hear or see. After an English class in high school, Sayers couldn't stop saying 'Macbeth'; later, he kept repeating the name of 1980s funk icon Rick James. 'There's so many different [movie] character names I would have said in the past.'

Not all repeated actions and habits are tics. 'It will only be classed as a tic if the person has the feeling they are unable to control it,' says neuropsychiatry professor Perminder Sachdev, a founder of the Tourette Syndrome Association of Australia.

Tics usually develop from age four, are more common in boys, get worse between ages 10 and 12, and ease off in adolescence.

Tics usually develop from age four, are more common in boys, get worse between ages 10 and 12, and mostly ease off in adolescence. One in eight school-children will have a tic for less than a year. 'It goes away [and] that's much more common,' says Valsamma Eapen, head of child and adolescent psychiatry at University of New South Wales. But tics last longer than a year for about one in 100 children. 'Often you will start with a blink or a facial twitch then it spreads to other parts,' says Eapen. Tourette's is diagnosed when a person has had at least one vocal tic and multiple motor tics for a year before they turn 18.

US pop star Billie Eilish, who was diagnosed with Tourette's when she was 11, has described her tics as 'very exhausting'. The intense focus of performing means she does not experience them on stage. Scottish singer Lewis Capaldi, diagnosed with Tourette's as an adult, has experienced tics during performances. At the Glastonbury Festival in 2023, tics forced him to stop singing mid-song, whereupon the crowd helped him finish the tune. 'I'm still learning to adjust to the impact of my Tourette's,' he said afterwards.

One complication of diagnosis is the fact that people might not display tics when they see a clinician. 'They wax and wane, new tics come all the time,' says Eapen. Doctors will quiz parents about their child's history or ask them to record videos. 'Just because it's not manifesting when we are consulting, or things are not there for a period of time, [we] should not discount it.'

WHAT PUTS YOU AT RISK OF TOURETTE'S?

Eapen once met a 13-year-old girl whose Tourette's had caused her to inappropriately touch her school teacher. But it wasn't the girl's tics that Eapen first noticed; it was her father's habit of raising his eyebrows. He had a 'kind of facial twitch', she recalls. 'And that was my clue that this was probably a tic family.'

Someone with a parent or sibling with Tourette's is 18 times more likely to develop the condition. Mandy Maysey's three sons all have Tourette's. 'I thought, *Oh my god, how did I manage to get three?*' says Maysey, president of Tourette Syndrome Association of Australia. 'But actually, when you realise it is genetic, those odds are an awful lot more likely.' Maysey has tics, too. 'If I'm around very ticcy people then I will tic, but I've never been diagnosed. My father definitely ticced.'

A small handful of genes are known to have a 'robust' link to the condition, says Professor Peristera Paschou, a leading researcher into the genetics of Tourette's at Purdue University in the United States. But they are just a tiny part of a bigger puzzle. 'It is totally feasible that we could be looking at 100 genes, of course each one of them contributing a small part of the risk.'

Between 50 and 60 per cent of people with Tourette's also have attention deficit hyperactivity disorder (ADHD) or obsessive-compulsive disorder (OCD). 'We have indications to show that the genes that underlie Tourette plus ADHD are different from Tourette plus OCD,' Paschou explains. James Leckman, professor of child psychiatry at Yale University, says people with tic-related obsessive-compulsive disorders will often arrange things in symmetrical order so they 'look or sound or feel just right' while those with ADHD have difficulty focusing. 'Almost all children I see have some degree of attentional difficulties, have some degree of obsessive-compulsive behaviours,' he says.

Genes involved in the immune system have also been linked to Tourette's. 'Of course, we don't yet know for sure,' Paschou says. Other factors could play a part: complications during pregnancy such as hypertension, infections, stress, pre-eclampsia and diabetes are linked to children developing more severe tics, several studies suggest. In people who are genetically predisposed, 'these stressors in pregnancy just increase that likelihood [of a child developing Tourette's] a little bit more,' says Professor Russell Dale, clinical director of Kids Neuroscience Centre in Sydney.

WHAT'S HAPPENING IN THE BRAIN WHEN SOMEONE TICS?

At the Johns Hopkins Tourette Center in the United States, co-director Harvey Singer uses the image of a race circuit to explain to families what could be happening in the brains of his patients, the car being a message, the circuit being networks in the brain. 'The problem could be at the start of the race or the quarter pole or down the finish,' he says. The car moves past the frontal cortex to areas such as the striatum, globus pallidus, thalamus, before racing back to the cortex but exactly where the car hits trouble 'we don't know for sure'.

The reality is much more complex: brain studies have shown that several neural pathways—a habitual behavioural circuit, a goal-directed circuit and an emotion-related circuit—could be involved. 'These pathways are fascinating but complex; it is not as simple as "this flows to that". They are multi-connected with inputs from lots of different structures,' Singer says. An imbalance of neurotransmitters, particularly dopamine, has also been linked to Tourette's, but whether it's a cause or symptom is unclear.

Most brain activity takes place behind a gate that keeps it from entering our conscious minds. But people with Tourette's, Eapen says, could be thought of as having a 'leaky gate' where 'a little bit of that comes to your conscious awareness'. This awareness, which people with Tourette's compare to the urge to itch or sneeze, is known as a premonitory sensation. 'You then respond to it with a motor action,' Eapen says. 'It's almost like the movements are released from the subcortical structures without the checks and balances.'

Yet she cautions against assuming that tics reveal what the person actually believes, even in cases where a tic happens in response to the environment. A person with Tourette's could call someone who is overweight 'fat', for example, but the impulse is acted upon before they're conscious of it. This is how Sayer explains the rare occasions when he blurts out something offensive: 'It's like a bad thought without bad intentions.' (Eapen prefers the term 'involuntary vocalisation': 'It is without thought or intention.') Our brains are not fully developed until our mid-twenties and tics ease off for many people because compensatory brain circuitry 'comes to the rescue' as they mature, says Eapen.

HOW IS TOURETTE'S TREATED?

The first tic Jack Van Hees noticed felt like a shiver. He was in primary school and his neck would jerk to the left when he was cold or when he felt socially uncomfortable. When he was a teenager, the tic intensified and he began repeating words such as 'drip' and swear words. 'I think there's a part in the brain where all of these funny words and bad words are stored [where] for some reason the Tourette's just loves to play,' says Van Hees.

Behavioural therapy eases symptoms in up to half of cases.

A psychologist introduced the then 16-year-old to behavioural therapy, which raises awareness of the premonitory urge to tic. 'It's training them to catch it,' Eapen says. Once aware of the urge, people focus on a competing action: contracting muscles at the back of the neck when there is an urge to move their jaw, squeezing their fingers when they feel like jerking their shoulder. Behavioural therapy eases symptoms in up to half of cases.

Van Hees says he was 'very unreceptive' to the treatment. 'They didn't know what it felt like,' he says. But the lessons sank in, and he taught himself to say 'blue square, red circle, yellow triangles' every time he felt the urge to curse. 'I can kind of move the energy somewhere else,' he says. 'Or it might even be a movement—I'll try to hold the noise and I'll turn that into a body tic.' The 21-year-old now has a tattoo of a blue square. 'It reminds me of where I came from.'

When tics severely disrupt the patient's life or cause pain, clinicians will offer medication that, Eapen says, can be effective in about 60 to 70 per cent of patients. 'It's more trial and error,' she says of finding a successful drug or dosage. Drugs used to improve attention and impulse control, such as clonidine and guanfacine, are often a first option, especially in children who also have ADHD. Clinicians can also prescribe antipsychotics, such as ones that target dopamine receptors, but they have a risk of greater side effects.

In rare cases the patient can receive deep brain stimulation (DBS) to regulate abnormal impulses. Surgeons implant electrodes controlled by a device similar to a pacemaker into certain parts of the brain. 'Obviously, you don't want to go there unless you really have failed all these other approaches,' says Singer. Sayers considered DBS but ultimately decided against it. '[Tourette's] is part of who I am; life can be hard, but it's what we make of it,' he says.

WHAT'S IT LIKE TO LIVE WITH TOURETTE'S?

When floods devastated the central Victorian town of Rochester in 2022, Caitlyn Quinn, a high-school student, lived with her family in a repurposed shed. The disruption to her routine made her tics worse, as did the lockdowns during the COVID-19 pandemic. 'I can barely sit at home on weekends. I have to be doing something,' she says. Her closest friend with Tourette's lives more than a two-hour drive away. 'I just wish that people would understand how people with Tourette's feel,' she says.

> 'I just wish that people would understand how people with Tourette's feel.'

She was bullied at school, which got worse after the floods. Her mother, Ingrid, believes the behaviour was a result of young people feeling out of whack. The bullies sometimes whistled, knowing full well it would trigger a tic in which Caitlyn punched herself. Like any teenager, she just wants to fit in. 'I have a few good friends who don't really care about it and just treat me like I am a normal kid.' She hopes to be a teacher or a doctor one day. 'I want to help other people with Tourette's, or just anything.'

Tics that cause people to harm themselves happen in only five per cent of cases. Kai Gardner, another teenager with Tourette's, has been hospitalised several times when involuntary movements caused injuries to her body. Bruises often appear where she has punched herself. Sometimes the tics cause her to hit other people in public. 'I always respond with, "Sorry" and "I can't help it. I have Tourette's. I can't control what I do",' Gardner says. In 2021–22, 42 people were admitted to hospitals in Australia as a result of Tourette's.

Some people with Tourette's develop a thick skin. 'If you worry about everyone's opinion all the time, it's going to drive you insane,' says Sayers. Gardner, who lives in western Melbourne, often worries about what others are thinking, especially if they laugh. 'If I found it funny, then, like, it's okay,' she says. 'If people are laughing anyway, I get self-conscious.' In a 2022 interview, Billie Eilish said the most common reaction to her tics was laughter. 'They think I'm, like, trying to be funny,' Eilish said. 'I'm always left incredibly offended by that.' In a later interview, she said she didn't want Tourette's to define who she was. Since publicly revealing her condition, she's 'learned that a lot of my fans have it, which made me feel kind of more at home'.

Usherwood is hyper-aware of others ticcing and the fact that he might start copying others' tics. 'Occasionally, that's led me to leave a room or a railway carriage,' he says. 'I was concerned that I might start to copy them and be thought to be making fun of them.' The Tourette Syndrome Association hosts camps around Australia where people with the condition and their families hear from experts and role models; do quiz nights, talent shows, ropes courses and rock climbing; and generally feel free from judgement. 'It means they can tic away, be themselves. They can be around other people who know exactly what's going on,' says Maysey, the association's president. 'It gets louder and louder and the tics compete.'

As with many people with Tourette's, Sayers has found his own way to temporarily control tics: drumming in a heavy metal band. 'If I do a show: start to finish, no tics,' he says. 'I have involuntary movements but as a drummer it's the exact opposite; I'm the one that keeps the time. I can't explain it, it's just that euphoric feeling.' Usherwood enjoys climbing for a similar reason. 'When I rock climb, I find the tics disappear but when I get to the top, I have to release them,' he says. Van Hees and Sayers, a gym coach, both enjoy climbing too. Directing attention away from tics might improve the signal-to-noise ratio in the motor system, says Eapen. 'Your attention is diverted away from executing anything else.'

Gardner plays tennis but on court the tics can get worse. At home, she enjoys writing fan fiction, and when she does the tics disappear: 'My imagination and my mind are on that. I get really involved.' In her first year of high school, she made posters of an iceberg to explain to students and teachers that tics are only the visible surface of life for someone with Tourette's. 'The top bit was Tourette's, the bottom was all the bits that come with it—ADHD, anxiety and it also said, "Bruises and pain",' Gardner says. 'I was the only one in school with actual Tourette's. And I felt a bit left out and that no one really understood.'

Gardner has several other neurodevelopmental disorders and is now home-schooled. She says this is the best option for her after having severe full-body tics at school. But before she left school, two boys asked her about the poster.

'They actually really wanted to understand what Tourette's is . . . They wanted to know more—and not just about what everyone expects.' As she tells the story, her eyes light up. 'And I told them a few things.'

11

CAN WE LEARN THE ART OF CONVERSATION?

Small talk is, in truth, a big deal.
Here's how to polish your party banter
and even get deep and meaningful.

Angus Holland

Fraser Lack is a chatty guy. As we speak on the phone he's constantly breaking off the conversation—always so politely it's not even vaguely annoying—to say hello to people he's bumping into on the street as he makes his way to the post office. 'Hi, Kerrie, how are ya? See ya!' Even the post-office worker gets his undivided attention for a few seconds: 'Thank you so much. Really appreciate it. Have a great day. Be well. See ya. Be well.'

A little later he laughs, 'In the last 19 minutes, I've said "hey" to that many people.' It's an impromptu demonstration of the topic of our conversation: how Lack, a real estate agent who has appeared on the reality TV show *Survivor*, became the kind of person who's comfortable talking to just about anyone. What, I wanted to know, can we learn from him and others blessed with the gift of the gab?

At some point, we will all find ourselves dragged to events where we'll be expected to make conversation with strangers, people we don't know very well or people we don't particularly like. Some you'll have little in common with. Others will hold forth with boring opinions. Still others will make zero effort to keep the chatter moving as you awkwardly attempt small talk.

So how can we get better at casual conversation? How can we draw others out of their shells? And how might we expand a few minutes with someone into a new friendship or a deeper, more meaningful relationship?

WHEN DID WE START TALKING ABOUT HOW WE TALK?

Over the centuries, some of our greatest thinkers have puzzled over the art—or science—of successful conversation, with varying results. 'What can be more delightful,' asked the Roman philosopher Marcus Tullius Cicero, 'than to have someone to whom you can say everything with the same absolute confidence as to yourself?'

Greek Stoic philosopher Epictetus came up with some unsurprisingly austere suggestions—'let silence be your general rule'—and warned against segueing lazily into what were, apparently, his era's 'common subjects', among them gladiators, horse races, athletes, and food and drink (although through today's lens, they sound pretty compelling—gladiators!).

Philip Stanhope, the fourth Earl of Chesterfield—he of sofa fame—is generally credited with coining the phrase 'small talk' around 1751 in a letter to his son. 'Study to acquire that fashionable kind of *small talk* or *chit-chat*, which prevails in all polite assemblies, and which, trifling as it may appear, is of use in mixed companies, and at table,' he advised.

Many 18th-century writers spoke of the 'pleasures (and pains) of conversation', noted Stephen Miller in his 2006 book *Conversation: A History of a Declining Art*. Jonathan Swift (author of *Gulliver's Travels*) wrote the satirical *Collection of Genteel and Ingenious Conversation According to the Most Polite Mode and Method Now Used at Court*. Samuel Johnson (he of dictionary fame) reckoned 'the happiest conversation is that of which nothing is distinctly remembered but a general effect of pleasing impression'.

In 1887, Irish scholar John Mahaffy published an early self-help guide to chit-chat. 'There can be no doubt that of all the accomplishments prized in modern society, that of being agreeable in conversation is the very first,' he ventured. 'Many men and many women owe the whole of a great success in life to this and nothing else.' Accordingly, he asked, 'Is there any method by which we can improve our conversation?' A product of the Victorian era, much of Mahaffy's advice concerns 'moral conditions' including sympathy, unselfishness, tact and modesty—although he warns against overdoing the latter. 'I need hardly insist that the man or woman who displays modesty by constantly apologising for native ignorance or stupidity injures conversation and can only amuse a company by becoming ridiculous.'

By the middle of the 20th century, such guides were endemic. Dorothy Draper's *Entertaining Is Fun!* suggested American women should regularly read one daily newspaper, one weekly magazine, one monthly magazine, one current novel and one non-fiction book, so they might have something to chat about. In 1961, British tastemaker and florist Constance Spry advised, in *Hostess*, 'Beware of talking too much of yourself and of what *you* think and do, avoid boring repetition, loud laughter, senseless giggles, affectation and interruption.'

One of the most successful self-help tomes is Dale Carnegie's *How to Win Friends and Influence People*, first published in 1936 and updated multiple times, selling more than 30 million copies to date. While some of his advice might appear a little dated, it remains an extraordinary work. As well as doing his own exhaustive research, Carnegie hired a researcher 'to spend one-and-a-half years in various libraries reading everything I missed'. They scoured countless biographies—including more than 100 books about former US president Theodore Roosevelt—and Carnegie interviewed 'scores of successful people', including the inventors Guglielmo Marconi and Thomas Edison.

'The rules we have set down here are not mere theories or guesswork,' he wrote. 'They work like magic.' We don't have room to set out all of his advice but here's one insight: 'Any fool can criticise, condemn and complain—and most fools do'.

While Stephen Miller observes 'there are far more books on improving one's sex life than on improving one's "conversation life"', the genre nevertheless seems to be flourishing. Among the recent offerings: Patrick King's *Better Small Talk: Talk to Anyone, Avoid Awkwardness, Generate Deep Conversations, and Make Real Friends*, Paula Marantz Cohen's essay '*Talking Cure*' (which we must credit for some of our historical touchpoints) and *Hello Stranger* by British anthropologist Will Buckingham.

> 'There are far more books on improving one's sex life than on improving one's "conversation life".'

'Interested in opening the door a little wider and finding ways of reconnecting,' Buckingham seeks out 'ways to free ourselves from the big, unwieldy problems of isolation and xenophobia . . . because if these problems can seem unmanageably vast, human ingenuity is bottomless, and sometimes the solutions we seek can be found in surprising places.'

WHY ARE SOME PEOPLE BETTER AT CONVERSATION THAN OTHERS?

Oscar Wilde 'was a spontaneous wit', writes Matthew Sturgis in *The Irish Times*, 'a happy conversationalist (with that gift for making his interlocutor feel almost equally brilliant) and a compelling storyteller.' It's a high bar. But being a genuinely engaged listener is probably more essential than having great anecdotes or lively opinions. 'I actually want to actively talk as little as possible,' says Fraser Lack. 'If I can leave the conversation having said nothing and learning everything about the person in front of me, that's a great conversation.'

As a Labor MP, Andrew Leigh is well practised at both talking with complete strangers and handling differences of opinion. 'A good conversation starts with wanting to learn from the other person—asking questions, keeping it light and being interested in the other person . . . just as a bad conversation is, essentially, a lecture which is ignoring the other person's views,' he says.

Another useful skill is the ability to understand who you're talking to, whether that's great aunt Edna or a dignitary from South-East Asia—something Nicholas Coppel had to largely learn on the job as a diplomat. When you're talking to foreign ministers one day and villagers the next, 'obviously, the content of the conversations can be very different, but also the style of delivery and the choice of language', he says. 'The key is to be able to understand your audience, understand their culture and the way they feel, and adapt your style to it. You can't really have a template or a rule book which tells you what to do. You need to make judgements about what's going to work and what isn't.'

Coppel, now an adjunct associate professor at Monash University, served as Australia's ambassador to Myanmar from 2015–18. In some countries, he notes, 'it's really important to let the other person speak. We [Australians] have a to-and-fro conversation style. You'll say something, I'll say something, we bounce off each other and sometimes interrupt each other to jump on a point. [Elsewhere], you have your turn, you speak, say everything you need to say. And then you leave it to the other person to have their turn, and they might speak for a long time as well.'

In diplomatic circles, a gift for small talk is helpful, particularly when you're starting out in a new country. 'The hardest part of a diplomatic assignment is those first few months,' Coppel says. 'You turn up at receptions, often in a ballroom of a five-star hotel with hundreds of people, you walk in, you know nobody. So you wander around looking for some other lost soul. But you just have to break through and make those connections.'

CAN YOU BECOME A BETTER CONVERSATIONALIST?

Rhetoric, the art of persuasive speaking, was taught in ancient times and enthusiasm for its classical concepts re-emerged during the Enlightenment in the 17th and 18th centuries. This training 'was tantamount to installing the operating system for adult social life', writes John Bowe, author of *I Have Something to Say: Mastering the Art of Public Speaking in an Age of Disconnection.*

Today, the art of small talk is 'fading', says Rupert Wesson, director of British coaching company and etiquette authority Debrett's. 'The point of small talk is that it allows rapport to be built before more serious subjects are tackled. Small talk might seem trivial, unimportant to some, but it is a way of making a personal connection with someone before tackling something that might be complex or even controversial.'

'A good conversation starts with wanting to learn from the other person.'

In the absence of formal schooling, you could probably do worse than delving into the many self-help books that promise to make you a Wilde-esque raconteur. In *How to Talk to Anyone*, for example, Leil Lowndes promises '92 little tricks for big success in relationships', among them: the 'killer compliment' ('What exquisite eyes you have!') and a technique called 'parroting'—repeating the last two or three words your companion says 'in a sympathetic, questioning tone'. Our favourite tip: 'Hello, old friend.' When meeting someone new, Lowndes writes, imagine they are an old friend with whom you're being reunited. 'The joyful experience starts a remarkable chain reaction in your body from the subconscious softening of your eyebrows to the positioning of your toes—and everything between.'

Donna Henson, an associate professor at Bond University, points to the 'social penetration' theory formulated by US psychologists Irwin Altman and Dalmas Taylor in the early 1970s. The theory frames how 'self-disclosure becomes our primary means of achieving intimacy with another human being', Henson says. 'So if we want to escalate a relationship in any way, it's really about talk, it's about revealing ourselves.'

'Small talk functions almost like an audition for friendship.'

Like Wesson she places great stock in idle chatter. 'I would say that small talk functions almost like an audition for friendship.' She advises avoiding 'conversational narcissism'—being overly concerned with yourself. 'A real conversation is a dialogue, almost without sides—that reciprocal give and take.'

And be positive, says Jessica Gopalan, director of marketing at training group Dale Carnegie Australia. 'What's our intention for the conversation?' she says. 'Do we want to be here? Because if we already think it's boring, we're probably not putting ourselves in the best space to be authentic. We need to show up, really putting ourselves in the other person's shoes.'

Ultimately, says Henson, developing conversation skills takes practice. Lack agrees. 'It is something that you have to make an effort in,' he says. 'It doesn't come naturally to everyone. It's something that's practised. I'll have a conversation with the postman. I'll have a conversation with the girl on the checkout at Coles. I'll be in line behind someone at the grocer around the corner from my place . . . and say, "I love that ice cream—great choice".'

OKAY, SO YOU'RE AT A PARTY. WHAT NOW?

Hopefully, your host, if there is one, has the nous to make thoughtful introductions: 'Angus, do you know Jerry? I believe he shares your interest in aardvarks.'

Going into a conversation cold is harder. Fraser Lack has the confidence to carry off a (sincere) compliment. 'I love menswear. I love clothing . . . and, you know, I love to leave people smiling or leave people feeling good about them-selves . . . A compliment is a really great place to start. "Hey, I love your shoes. Where did you get them from?"' Or he might mention something odd he learned from a podcast. 'That's a really good starting point,' Lack says.

But what do you do if you're at a gathering and want to connect with someone who is already holed up in a group? Coppel says that, whenever he found himself in this situa-tion, he would stroll over and gently touch the person on the elbow. 'That will usually cause somebody to turn to you,' he says. You can then say, 'I've been meaning to talk to you.'

Once you have a foot in the door, it's question time. In 2017, a team led by Karen Huang at Harvard University identified 'a robust and consistent relationship between question-asking and liking'. They concluded: 'People spend most of their time during conversations talking about their own viewpoints and tend to self-promote when meeting people for the first time. In contrast, high question-askers—those that probe for information from others—are perceived as more responsive and are better liked.'

Questions can keep a conversation rolling—as long as they're not intrusive, overly interrogative or what the

Harvard researchers described as simply 'rude'. Keep your queries open-ended, suggests Jodie Bache-McLean, managing director of the June Dally-Watkins organisation, founded in 1950 by the woman known as Australia's Queen of Etiquette. 'That's conversation 101—people will have to elaborate and not give just a yes or no answer.' For example, you could ask, 'What do you spend most of your time doing?' rather than, 'What do you do for a crust?' Don't fire off 20 questions in a row either, says Bache-McLean—it's a conversation, not a Q&A session.

Before wading in, especially with people you don't know well, you might seek permission to ask a question. 'Such as, "Would it be okay if I asked you about . . .?" and make it a genuine question,' suggests Wesson. 'We Brits use this sort of indirect language quite a lot. Australians, at the risk of generalisation, tend to have a more direct style, but the principle is still a good one.'

Don't ask about somebody's family unless you know them a little already—they might be recently bereaved or divorced and feel uncomfortable talking about it. 'I don't have children, so that sometimes invites an awkward conversation,' says Henson. 'If people ask you, "Do you have kids?", okay, well, what are my options here? At least in anticipating it, I'm prepared for that possibility.' In a similar vein, says Bache-McLean, 'Be mindful of giving unsolicited advice.'

And when you're asked the inevitable dull question, such as 'Where are you from?', never give a one-sentence response, writes Lowndes. 'Give the hungry communicator something to conversationally nibble on. All it takes is an extra sentence or two about your city—some interesting fact, some witty observation—to hook the asker into the conversation.'

HOW IMPORTANT IS BODY LANGUAGE?

'I definitely make a conscious effort with how I present myself,' says Lack, 'so, in terms of using very positive, open and inviting, welcoming body language, right, rather than being closed off. Negative stance, crossing your arms—these are things that I never do. You have to make an effort but, slowly, it just becomes learned behaviour.' It's the same with smiling. 'A couple of years ago, I realised that I wasn't a smiler,' he says. 'I realised I had quite a stern face. I actually had to make an effort to smile.'

We've all encountered somebody we're trying to talk to peering over our shoulder or around the room, perhaps hoping to find somebody more interesting than we are. 'We talk about these things as though they're separate, what we say and our body movement, our non-verbals,' Henson says. 'But really, in most situations, it's this holistic package of signals.'

We've all encountered somebody we're trying to talk to peering over our shoulder ... hoping to find somebody more interesting.

In its conversational skills ratings scale, the US National Communication Association scores talkers on 25 elements, including speaking rate, vocal confidence and volume. The list also has nine non-verbal cues, such as posture, 'shaking or nervous twitches' and 'nodding of head'.

Most people clench their fists, fidget or slouch when they're bored or stressed. Open palms and relaxed shoulders show the other person that you're engaged, making them feel respected and confident. Avoid the mixed signal. 'The subject I'm talking about, the words are all positive but if I've got my arms folded, eyes looking down—you're probably not going to trust what I'm saying,' says Gopalan.

'Our body language needs to match our message. We need to be congruent and not ignore the impact of our body language.'

While there's a whole science behind non-verbal communication, Wesson boils it down to this: 'All that is needed is to consciously relax (our hands, our feet, our face), make sure we smile and don't forget eye contact. Put simply, it turns out that by doing these things we put ourselves and others at ease.'

SHOULD YOU REHEARSE AHEAD OF A SOCIAL EVENT?

At parties and other gatherings, we're all asked the same inevitable questions: 'How are things?' 'How's work?' 'How's the family?' Or, during the festive season, 'Glad to see the back of that year, eh?' Should you work on your answers ahead of time?

Probably not, says Lack. 'I honestly just try to be myself in absolutely everything I do. Because no one can keep up the facade for too long. It's exhausting. And you will burn out. It's inauthentic and people will see through that. Someone might leave a conversation or engagement with someone and go, "I don't know what it was. I can't put my finger on it. But something about that person just isn't right."' Bache-McLean is also wary. 'People are really quick to pick up on pre-rehearsed dialogue.'

Dale Carnegie's Gopalan says a repertoire of small talk can be useful sometimes, 'particularly where we're in conversations or situations with people that we're not very familiar with, people that we don't know that well, maybe people we've had a conflict with in the past. We may be forced to sit around a table for a couple of hours with people we don't know or have that much in common with. So having in the back of our heads a framework, or some planned questions or topic ideas can really give us that kind of ease and security going into it.'

Dale Carnegie noted that the US president he researched so much took things even further: 'Everyone who was ever a guest of Theodore Roosevelt was astonished at the range and diversity of his knowledge. Whether his visitor was a cowboy or a Rough Rider, a New York politician or a diplomat, Roosevelt knew what to say. And how was it done? The answer was simple. Whenever Roosevelt expected a visitor, he sat up late the night before, reading up on the subject in which he knew his guest was particularly interested.'

What should you do when the conversation flags or—worse—takes an unpleasant turn when, say, someone expresses a strong and unwelcome opinion? 'In figuring out how to live together, we have to negotiate differences,' says Andrew Leigh. 'So when you're speaking with someone who disagrees with you, that's a good thing. That's what I remind myself when I'm chatting with a friend or a relative who disagrees. It's about negotiating difference. Or there may be instances in which you choose not to have that conversation and then set some clear boundaries around areas you're willing to explore and areas you're not willing to explore.'

Leigh suggests that instead of taking issue with some-body's views, try to find out more about why they hold them, 'whether you're talking about the Middle East or whether you're talking about economic policy: "Why do you hold that view here? Tell me a little bit more about it. Tell me about how you came to those views." And then you can move beyond the sort of shouting points that might be downloaded from the internet into getting into something a little bit more personal.'

If it gets too much, try a simple, 'I might leave that for another time', says Bache-McLean. And if, despite your best efforts, the conversation keeps falling flat, how do you exit gracefully? Bache-McLean suggests: 'I'm going to go and chat to the host to see if she needs any help in the kitchen.' And if you're sitting next to somebody at a table, says Coppel, 'There's no point making them feel more uncomfortable and awkward. And there's always a person on either side and there are people across the table. You have options.' At a stand-up affair, he says, you can always use the old standby, 'Can I get you a drink?'. 'And you go off and never come back. It's a big room, you could get lost on the way back. It's not rude. It's understandable.'

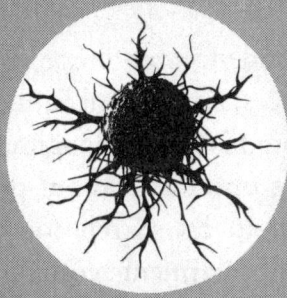

12

WHY IS CANCER SO HARD TO CURE?

Despite breakthroughs in treatments, cancer remains one of the world's biggest killers. What makes a cure so elusive?

Kate Aubusson, Jackson Graham and Felicity Lewis

A crisis of confidence can come from the most unassuming sources. For Darren Saunders, it was a simple question from his young daughter. The cancer biologist had been feeling optimistic about the progress medical science was making in its search for cancer cures, with 'whole new classes of treatments coming through'.

'I was going around telling everybody about all these amazing things we can do and the advancements we've made in understanding cancer,' Saunders recalls. 'Then my kids' grandmother got lung cancer and I couldn't do anything to save her. My daughter asked me, "Why can't you just go to the lab and mix up some medicine to cure Nonna?" It was a real slap in the face. It made me contemplate for a long time what I was doing.'

Why haven't we cured cancer? After all, the cliché goes, if we can fly to the Moon . . . We've found 'cures' for other deadly diseases. As COVID-19 swept the world, scientists created vaccines at breakneck speed. Yet, despite decades of research costing hundreds of billions of dollars, cancer is still one of the world's biggest killers, responsible for millions of deaths every year; in Australia, 50,000 people succumb to some form of cancer each year.

What is cancer? What makes it so hard to 'cure'? And what are some of the latest treatments?

WHAT IS CANCER?

When he is asked why we haven't cured cancer, Professor Glen Boyle doesn't hesitate: 'Because it's bloody hard!'

To start, cancer isn't a single disease; it's a constellation of more than 200 diseases. It has been linked to mutations in between 500 and 1000 genes (of the 20,000 or so genes in each person). 'We are literally trying to tackle hundreds of different diseases and none are easy,' says Boyle, head of the Cancer Drug Mechanisms Group at the QIMR Berghofer Medical Research Institute in Brisbane. 'Every cancer is different—so there's our first problem. In breast cancer alone, there are 20 to 30 different cancers [differentiated by their genetic muta-

Cancer isn't a single disease; it's a constellation of more than 200 diseases.

tions]. Almost every patient has a different kind of disease that behaves differently in their individual body.'

This is where the cliché about the Moon breaks down. 'We know what the Moon is, we know where it is, we know how far it is, what it's made of—so it's a fixed target,' says Saunders. 'Cancer is very different in that we still don't completely understand what it is and how it works. In terms of that metaphor, it would be like trying to fly to thousands of different moons in thousands of different places . . . and halfway to a moon, it might change shape or direction, or move into a different part of the sky.'

For a long time, cancers were categorised by the organ in which they first appeared but today, cancer subtypes are increasingly being described by their genetic or molecular markers. In breast cancer, for example, we might talk about the mutations in the BRCA1 or BRCA2 genes. Or we might talk about HER2-positive cancer, which tests positive for a protein called 'human epidermal growth factors receptor 2' and tends to be more aggressive than other types of breast cancer. These markers 'allow us to better define the thing we're working against,' says Saunders, now the deputy chief scientist and engineer for the New South Wales government.

Scientists can now examine a tumour's 'fingerprint', says Boyle. 'We can see up to six different sorts of melanoma cells in one tumour and they are all doing different things, mutating and growing at different rates and potentially responding differently to potential treatments.' When he started working on melanoma, the consensus was that one person's melanoma was the same as another's. 'But when we looked at the genome, we realised, well actually, no, it's not the case at all.'

Still, all cancers do have something in common: they're made up of cells that have lost the normal ability to control their growth.

Cancer strikes when a cell ignores the body's inbuilt controls and multiplies uncontrollably.

The trillions of cells in our bodies have specific functions—liver cells to break down fats and produce energy, red blood cells to transport oxygen around the body—and follow a predictable cycle of growth, division and death. 'We have all of these complex control mechanisms that make our cells copy themselves only when they need to,' says Saunders.

Cancer strikes when a cell ignores the body's inbuilt controls and multiplies uncontrollably. In some cases these haywire cells can travel from the original site of the cancer through the blood or lymphatic systems, burrowing into tissue, where they flourish. 'They become really nasty and can invade and spread into other parts of the body,' says Saunders, 'and that is when you have very serious and often untreatable versions of the disease.'

A cancer that spreads is called metastatic. Medicos call the original tumour 'primary' and the secondary tumours 'mets', short for metastases. Pathologists speak of grades of cancer (how aggressive it is) and stages of its development (how much it has grown and how far it has spread). Although the circumstances are different in each case, generally a small tumour that has not spread is described as Stage 1, while cancer that has spread to at least one other organ is Stage 4.

Common cancers found in parts of the body not essential for survival (say, mammary glands in the breast) only become potentially deadly when they infiltrate vital organs such as the lungs. 'Let's say you have a massive ball of cancerous cells growing in your lungs,' Saunders says. 'That part of your lung is no longer going to be working as a lung should because those cells have lost their ability to do their jobs.' Cancer can spread to the liver, stopping it from keeping toxins out of our system. It can cause blood to become 'stickier', heightening the risks of blood clots in the brain, where a stroke can occur, or in the heart. Tumours can push on parts of the brain, for example, eventually causing a person to lose alertness and then consciousness. With their immune system weakened, a patient can be susceptible to pneumonia, say, or influenza, which can contribute to their death.

If cancer in one of his patients has not yet spread, Associate Professor Tom John will use the word 'cure' when speaking with them 'from the get-go because the whole intent of treatment is curative'. But with patients with metastatic cancer, John, who is deputy director of oncology at the Peter MacCallum Cancer Centre, is more circumspect, telling them that 'we're aiming to control this, but we can't cure it'. However, some new treatments are helping people recover from metastatic cancer (more on this later). If, over a long period of time, scans continue to show no signs of cancer, John will say: 'We don't use the c-word very often but, in this case, you may well be cured of the cancer. We don't know this for sure. But, at this point in time, it certainly looks that way.'

WHY CAN'T WE JUST STOP CANCER CELLS SPREADING?

The fact that cancer is born from our own cells—rather than from an invading foreign infection—makes it particularly tricky to treat. It's far easier, although still an incredible feat, to produce an antiviral vaccine because the structure of a virus is materially different from that of a human cell.

'You can produce a vaccine against something like the spike protein of the coronavirus because human cells don't have spike proteins, so it'll kill the virus but won't kill the cells around it,' says Ian Olver, adjunct professor at the University of Adelaide's School of Public Health and one of Australia's leading cancer researchers, bioethicists and medical oncologists. 'Cancer cells are far too much like the rest of the body's cells, so you have got to find something on the cancer cells to target with some sort of agent that isn't on the normal cells and so won't harm them. Otherwise, you've just got something that's toxic to both.'

Our innate immune system can be very effective in targeting cancer cells and leaving normal cells unscathed. 'It's a finely tuned, super-sensitive system that is able to recognise things that don't look like our cells and attack and kill them off,' says Saunders. 'We would have a lot more cancers than we currently do if we didn't have functional immune systems.'

Some breast cancers communicate with their micro-environment to camouflage themselves or disarm the immune response.

The trouble is that cancers evade and suppress the immune system. Some breast cancers, for instance, communicate with their micro-environment—immune cells and blood vessels—to camouflage themselves or disarm the immune response, explains Associate Professor Pilar Blancafort, program head of cancer epigenetics at the Harry Perkins Institute of Medical Research in Perth. 'Triple-negative tumours can send chemical messages via small molecules called chemokines that talk to the immune cells (T cells) and deactivate them,' she says.

In this way, they can also develop resistance to promising treatments. 'Some aggressive tumours have so many types of mutations in their genetic material that make them primed for rapid evolution, which enables them to spread quickly to other parts of the body,' says Blancafort. 'The new emerging clones may be able to survive therapies and become resistant to them. Then we've got to find alternative treatments. We are currently targeting the genes in the cancer cells that are responsible for these "talks" with the immune system.'

There are other curveballs. In pancreatic cancer, for example, non-cancer cells can account for up to 80 per cent of a tumour. They can form a rogue organ—complete with its own blood supply, nerves, lymphatic system and so on—that enables the cancer cells to grow while shielding them from the body's immune system, which is why symptoms might not be felt until the cancer is well advanced. Pancreatic cancer has one of the worst survival rates of any major cancer worldwide. In 2022, the Australian five-year survival rate was 12.2 per cent; just over a third of people will still be alive one year after diagnosis.

Saunders calls cancer a perfect storm. 'They are horrible little things, but this amazing, adaptive behaviour is one of the things that make them so fascinating to study.'

This adaptability means that killing 99.9 per cent of the cells in a tumour may well not be a cure. Surgeons, for example, can remove a tumour and follow-up treatments can 'mop up' remaining cancer cells. But leaving just 0.1 per cent of the tumour potentially gives the cancer wiggle room to take off again. 'Unless you kill 100 per cent of the cells,' says Boyle, 'that tiny leftover remnant has already experienced that therapy agent and can adapt to basically stop itself from being killed off by another whack of that therapy.'

This is why cancer patients can feel 'scanxiety' as they await test results after treatment. Boyle has firsthand experience. 'I almost drove myself insane when my then fiancée was diagnosed with breast cancer six weeks before we were due to get married,' he says of his wife Sarah, a fellow cancer researcher. 'I had just started doing a little bit of breast cancer research and, knowing what I knew, I dove deep into analysing her case and prognosis.

'In the end, I had to step away and let her oncologists do what needed to be done. The good news is she's now my wife, we have two kids and after almost eight months of chemotherapy and seven years of anti-oestrogen therapy, we count ourselves very, very lucky.'

CAN THE ROOT CAUSE OF CANCERS BE TARGETED?

It sounds like a great idea, but we don't yet completely understand what causes cancer. 'They've all got different starting points,' Boyle says. 'Take melanoma. We know that there is a very specific cause for the vast majority of these cancers, which is UV exposure, and we know that in about 50 per cent of these cases, the damage will be to a specific gene called BRAF.' BRAF melanoma can be treated with therapies that attack the BRAF protein to shrink and slow the growth of tumours. But in another 25 per cent of melanoma cases, the UV damage hits the NRAS gene, triggering more aggressive mutations notoriously known as 'undruggable'.

A 2017 study of more than 7500 tumours across 29 cancer types discovered it takes just a handful of specific mutations (between one and ten) to convert a normal cell to a cancer cell. It also found that the number of mutations driving a cancer varies depending on the cancer type. For instance, about four mutations on average lead to liver cancer, but colorectal cancers need roughly 10 mutations to emerge.

It takes just a handful of specific mutations to convert a normal cell to a cancer cell.

'Say a cancer requires five specific mutations and you have four of them. Then you're in the clear,' explains Olver. 'But your next-door neighbour is unlucky [enough] to be hit by all five mutations, and that will often trigger the cancer.' Sometimes you can inherit a couple of the mutations and a couple more will come along through exposure to your environment. The risk that mutations pose plays out over a lifetime, Olver adds. If you carry mutations associated with a particular cancer, 'it doesn't mean you'll get it next week'.

The truth is that the genesis of cancer is often a horrible mix of genetics, environmental factors and bad luck in the guise of a random mutation in a person's genes, Saunders says. 'Once you've been dealt this bad hand, fixing the problem is like trying to unscramble an egg.'

Age is the single biggest risk factor for cancer. Just over 100 deaths from cancer were forecast in 2023 for every 100,000 people aged 40–59; among 60- to 74-year-olds the number rose to 388 deaths; and for people aged 75–89, it was 1178. 'As we age, cells have had more opportunity to divide and multiply and therefore more possibility of incorporating mutations into the genes as they divide,' says Saunders. To offer some perspective, it's estimated that around 43 per cent of Australians will be diagnosed with cancer, not including the most common skin cancers, by the time they get to 85. A man has a 17 per cent risk of dying from cancer by this age and a woman 13 per cent, according to the Australian Institute of Health and Welfare.

WHAT ARE THE PROMISING NEW TREATMENTS?

When doctors found a tumour on Lisa Govelli's left lung in 2019, it had grown so large it was pushing against nerves in her spine, causing her left eye to shut and triggering pain all over her body when she did simple tasks such as making lunch for her two kids. The 36-year-old was quickly admitted to Peter MacCallum Cancer Centre to receive intense radiation therapy six days a week. 'With radiation, they had to shrink it away from my spine straight away because they said if it did grow any more that I would probably risk paralysis,' she says.

The cancer had spread to five other sites across Govelli's torso so she began chemotherapy too—the most common cancer treatment, in which chemicals kill cancer cells (and, inevitably, some healthy cells).

The chemotherapy seemed to have almost cleared Govelli of cancer when more bad news came: another tumour was growing on her spine. 'For that to happen again, I was sick in the stomach,' she says. By this time, she had heard about a breakthrough treatment: immunotherapy, which turbocharges the body's own immune system to fight cancer. Months after starting the treatment, the new tumour had disappeared and by 2024 she was considering whether to stop the drug. 'Everything has been clear for so long now,' she tells us. 'That's given me confidence. Eventually, you have to stop staying in the fear of the original diagnosis.'

Immunotherapy is a paradigm shift from the model of using chemicals to kill cancerous cells, says Olver. 'This is where we are putting our hopes for finding cancer cures.' For decades, doctors had been trying to harness the immune system to combat cancer. Then scientists made a Nobel-prize-winning discovery. It turns out our immune cells, although designed to attack abnormal cells, also carry proteins that act as checkpoints, to stop them going into overdrive and killing healthy cells too. By reacting with these checkpoint proteins, cancer cells can disarm immune cells. But when scientists worked out that the checkpoints can be inhibited—the brakes on an attack can be removed—they ushered in a new type of cancer drug called immune checkpoint inhibitors. '[They] take away the stop signal and allow the immune system to then recognise the cancer cell as abnormal,' says Kate Mahon, director of medical oncology at Chris O'Brien Lifehouse in Sydney. Or, as Saunders puts it, 'They basically wake up the immune system and say, "Hey, there's a tumour over here. Come and take care of this".'

Another type of immunotherapy involves scientists genetically modifying a patient's own immune cells in a lab so the cells are better equipped to recognise and fight cancer, then infusing them into the patient. CAR T cell therapy is so named because it adds 'chimeric antigen receptors' to T cells, which are immune cells that are particularly effective at fighting cancer. It's used to treat some blood cancers in Australia. Cancer antigens vary so the therapy can be targeted to a specific person's cancer—personalised cell therapy.

Immunotherapies are no silver bullet, though. Even among people with the same type of cancer, the drugs might work in some but not in others. They also have side effects (some so severe the treatment has to be stopped) and not all attract government subsidies in Australia, making some of them very expensive.

One immunotherapy drug that is subsidised is the checkpoint inhibitor Keytruda, which targets a protein called PD-1. It is on the Pharmaceutical Benefits Scheme for Australians with certain types of Hodgkin's lymphoma, melanoma and lung, bladder, colorectal, head and neck squamous cell, kidney and endometrial cancers. In some cases, it has had remarkable results. In a 2021 study, a third of lung cancer patients had high expression of the PD-L1 protein, and a third of them were alive five years after receiving Keytruda. 'It's incredible,' says John. 'We've been able to discharge patients. [They're] people who came to us with Stage 4 disease that has spread throughout their body and would normally have been dead within 12 months.'

When Charmaine Blanch felt what she thought was ligament pain, she put it down to discomfort brought on by her pregnancy with her daughter, Florence. Her own mother, Anne, had died of bowel cancer, aged 53, seven years earlier. 'But you still don't think you're ever going to get it,' Blanch says. After the shock discovery she had the cancer herself, she underwent surgery to remove a portion of her bowel. 'After that surgery, I was like, I don't want to go back on the table,' she recalls. She was told she had two to three years to live.

Her mother had undergone chemotherapy, but Blanch wanted to avoid the side effects and stay as healthy as possible. 'I was offered chemotherapy twice and said no,' she says. 'In case I wasn't there too long . . . I wanted my kids to see me at my best.' Keytruda wasn't yet on the PBS for the cancer that Blanch had (which makes up 5 per cent of metastatic colon cancers) but as Florence marked her first birthday, Blanch received access to the drug at the full cost

cost of $80,000 over several years. In 2023, three years after the initial diagnosis, Mahon, who treated her, told Blanch she was cancer free. 'You don't get to give that level of good news very often,' Mahon says. 'Fingers crossed, that's it.'

A weight has certainly lifted for Blanch. 'Cancer gives you a new meaning to life—all the little stuff doesn't matter anymore,' she says. The improvements in cancer treatments in just a few years have been dramatic. Blanch wonders if the outcome would have been different for her mother today. 'I just wish my mum had had access to immunotherapy.'

Timing is just one of the poignant aspects of cancer breakthroughs. 'With every new therapy, there's a wave of enormous enthusiasm that gets pared back over time,' says Olver. 'Immunotherapy is a genuine seismic event but not the universal cure with no side effects that some people were hoping it would be.'

'Immunotherapy is a genuine seismic event but not the universal cure ... that some people were hoping it would be.'

Another emerging area of treatment is in targeted therapies, which aim to disrupt the molecules that enable a cancer to multiply by targeting specific genetic mutations or proteins. In most cases, doctors diagnose cancer by taking a sample of the tumour, which can then be used to sequence the person's DNA. 'We look for a specific number of mutations and if we find them, it enables us to use a specific therapy that targets that particular type of cancer,' says John. Targeted therapies on their own tend not to cure people with metastatic or advanced cancer. However, they can help clear cancer cells left after surgery and chemotherapy to increase the chance of cure. In some studies, they have been shown to reduce deaths by half in people in the early stages of a particular type of lung cancer.

The hope is that one day scientists will be able to pinpoint exactly what drug will work for any given individual. 'We can say in many hundreds of people this percentage is going to benefit from the drug, but you don't know if the particular patient you've got in front of you is going to be [among] them,' Mahon says.

Meanwhile, we have made progress with some preventative measures. Cervical cancer is one of the most preventable and treatable forms of cancer, thanks to the HPV vaccine and screening. Some 90 per cent of bowel cancer cases can be treated if caught early. 'It's possible to consider eradicating cancer of the bowel if we can get bowel-cancer screening rates up,' Olver says. Australians receive a bowel-screening kit in the mail every two years from age 50 to 74, and as of July 2024 those aged from 45 can request a kit online or through their GP. But of the 6 million people sent kits in 2021–22, only about 2.5 million returned their samples. The beauty of bowel cancer screening, explains Olver, 'is that [any] bleeding is due to the presence of a pre-cancerous polyp [that can be] treated before it develops into invasive cancer.'

Overall, a diet high in vegetables and fruit and low in high-calorie carbohydrates, sugars and red meat, combined with regular exercise, limiting alcohol intake, not smoking and protecting skin from sun damage can reduce your risk of cancer, Olver says, even if nothing can guarantee you won't ever get it. 'If universally we did all these things, we would reduce the incidence of cancer quite a bit and we would need to rely far less on new discoveries.'

'People are always going to get cancer ... But I think we're moving towards a situation where it's a manageable disease.'

For Glen Boyle, watching cancer treatments emerge over his career has been a 'huge rollercoaster'. Some drugs that had offered new hope began to fail in some patients as cancer cells became resistant to them. Then other drugs, such as the checkpoint inhibitors, were discovered. Today, Boyle is optimistic. 'Before 2011, there was less than 10 per cent survival from Stage 4 melanoma, and in 2021 it was greater than 50 per cent. That's an incredible accomplishment,' he says. What of the future? 'People are always going to get cancer. Just by nature, it's always going to happen. 'It happens in all animals—dogs get cancer, cats get cancer. We're never not going to be with it. But I think we're moving towards a situation where it's a manageable disease. And I think that's a great place to be.'

13

COULD WE EVER JOURNEY TO THE CENTRE OF THE EARTH?

The deepest hole in the world goes down
12 kilometres—still a long way from our
planet's core. What would we find
if we kept digging?

Sherryn Groch

Y ou are standing above a churning sea of molten metal hundreds of kilometres deep. Fortunately that's not all you're standing on, there's some rock too—a lot of it, actually. And crystals. More than a few dinosaur bones as well. And deep under all that, in the very heart of our blue planet, burns a core as hot as the surface of the sun.

In his 1864 novel *Journey to the Centre of the Earth,* Jules Verne imagined a lost cavern of dinosaurs, giant crystals and a sprawling sea beneath our feet. The truth is just as strange. 'It's a planet within a planet,' says Professor Hrvoje Tkalcic, a geophysicist at the Australian National University in Canberra. Within the Earth is a core of crystallised iron almost as big as the Moon. It has its own spin and structure, distinct from the rest of the planet, but generates an electromagnetic field that makes life on the surface possible. 'And, if we could see it, it'd be like looking at the sun,' says Tkalcic.

There really are crystals and even a huge 'ocean' of sorts hidden inside the Earth.

Only in the past century have we even known the core existed. Because of the extreme temperatures and pressures beneath the surface of our planet, it's harder to get to than outer space. We probably know more about the rest of the solar system than Earth's own interior.

But Verne's imaginings weren't all fanciful; there really are crystals and even a huge 'ocean' of sorts hidden inside the Earth. Now, as an ambitious drilling project gathers steam and laboratories come closer to simulating the punishing conditions of the core, scientists hope to crack even more of the planet's secrets.

What's the Earth made of? Is it really a giant magnet? Could we ever journey to its centre?

HOW DO WE KNOW WHAT'S INSIDE THE EARTH?

The Earth formed, with the rest of the rocky objects in our solar system, out of the dust and gas left over from the making of our sun some 4.5 billion years ago. Originally a molten ball of fire, it cooled enough to get a crust of rock on its surface while heavy elements, such as iron and nickel, sank deeper into its gut to forge the core.

But it's plain that Earth is also different to everything else we see in the solar system: ablaze with life, it has an engine at its heart that's more than just the dying fire of its formation. This unique 'engine' was formed back at the start. As astrophysicist Professor Jonti Horner says, Earth may be one of the luckiest planets in the universe. Even a cataclysmic early collision with Theia, a planet the size of Mars, helped foster the right conditions for life (and forged our stabilising moon from the debris).

We can't see into the Earth and, so far, deep drilling projects haven't worked. But we do know that when our planet is rocked by big earthquakes, it vibrates like a bell as shock waves bounce around the interior. Just as X-rays and soundwaves can reveal the shape of whatever they pass through, these 'seismic waves' give us a window into the Earth's interior, explains Professor Arwen Deuss, a geophysicist at Utrecht University in the Netherlands. 'It's like a light going on.' The denser the material, the faster seismic waves will travel through it, allowing scientists to gauge the properties within.

You've likely seen the four layers of the Earth in school textbooks, in neat circles of red and yellow. We live on the crust, the thin, rocky skin of our planet. It's broken up into the slow-drifting tectonic plates that make and remake the continents. The crust runs only about 35 kilometres deep on average (it's thickest beneath mountain chains and thinnest on the ocean floor); it's another 6000 kilometres to the very centre. 'The crust [accounts for] just 1 per cent of the planet's mass,' says Horner. 'We're like bacteria clinging to the outside of an apple.'

In 1909, Croatian geophysicist Andrija Mohorovicic noticed that waves on seismographs suddenly sped up beyond the crust. This led to his discovery of Earth's next layer: the mantle (its boundary with the crust is now known as the 'Moho' discontinuity, in his honour). The mantle, nearly 3000 kilometres thick, is largely made of a rock called olivine, giving it a green colour, but it has red flecks of garnet throughout, Deuss says. And it's also on the move. Tkalcic compares it to fingernails growing as the tectonic plates shift overhead, pulling up new crust to the surface and pushing down the old as part of the planet's vital carbon cycle. 'Think of it as a conveyor belt,' he says. 'Or a hot soup on the boil where things rise to the surface and others sink. And at the very bottom is a graveyard.'

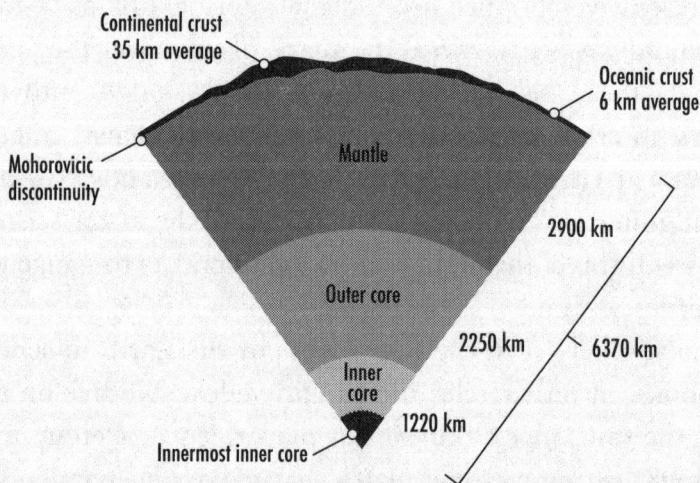

Continental crust
35 km average

Oceanic crust
6 km average

Mantle

Mohorovicic
discontinuity

2900 km

Outer core

2250 km

6370 km

Inner
core

1220 km

Innermost inner core

Inside the Earth. *Simon Rattray*

Ancient slabs of crust and tectonic plates lie broken where the mantle ends, alongside two mysterious continent-sized blobs that straddle the core like a pair of earmuffs. No one has figured out what these 'mantle anchors' are, says Deuss, but we know their density and chemistry are different to that of the surrounding rock (thanks not only to seismic waves but also to tantalising samples hurtled to the surface via volcanoes). According to one theory, they are the remains of that planet Theia, which may have fused with the young Earth or been swallowed whole during the ancient collision.

Next, just over halfway into the Earth, you hit the outer core—a sea of liquid metal burning at thousands of degrees Celsius. Its swirling motion generates an electric current and creates the Earth's magnetic field. Further in, about 2500 kilometres wide, is the inner core where the downward force of the planet is so great that the melted metal is compressed into a solid state. 'Here the pressure takes over . . . like a neutron star,' Tkalcic says. 'The atoms are so tightly packed in. Imagine an atmosphere more than 3.5 million times heavier than at the surface.'

Danish seismologist Inge Lehmann discovered the inner core in 1936 when she noticed seismic waves refracting off its boundary. Deuss and Tkalcic have since confirmed, separately, that the inner shell is solid by spotting the presence there of special waves, known as shear waves, which travel only through solids. In fact, this iron ball has been growing as the Earth slowly cools, eating into the liquid outer core by about one millimetre every year.

Scientists have learned about the components of planets by studying asteroids and meteorites, themselves often the broken cores of ancient worlds lost in cosmic collisions, says Horner. The density of Earth's core indicates that it's likely made of iron and nickel with some lighter elements, such as oxygen, silica and sulfur, mixed in. Because of how fast iron conducts heat, they can tell the inner core is younger than the Earth from its size.

Best estimates suggest that the inner core started to solidify anywhere between 1.5 billion years ago, around the time continents started to form, and 500 million years ago, during a period when life was rapidly diversifying across the globe, known as the Cambrian explosion or the Biological Bang.

At the very centre of the Earth is a fifth layer—an innermost core that was only recently discovered, 'like a pip in a piece of fruit', Deuss says. It doesn't have as clear a boundary as the other layers, which can reflect seismic waves like a mirror. If it did, it would have been discovered earlier, Tkalcic says. 'Now we need to change those school textbooks and make four layers five.' Scientists think the iron crystals in this very innermost core are orientated differently to those in the rest of the inner core, the reason for the difference still a mystery. While the inner core's crystals are growing at about a millimetre a year, the innermost ones are fossilised. 'The crystallisation started in the centre,' he says, which is the oldest part of the inner core. 'It's like a time capsule.'

Deuss and her team at the University of Utrecht recently created a 3D image of the Earth's elusive inner core by piecing together decades of earthquake data—a feat that was only possible with the advent of supercomputers. Some of the best data about the Earth's interior come from earthquakes near the volatile South Sandwich Islands in the South Atlantic Ocean. They ring almost straight up through the planet's interior to ping seismometers on the other side, in Alaska. Recordings of the shockwaves from nuclear bomb detonations, such as Cold War tests, provide even finer detail. 'But they didn't travel as far,' Deuss says. 'They're too high-pitched. And we need more seismographs at sea too. There's whole stretches of the planet we're missing.'

HOW DEEP HAVE WE DRILLED INTO EARTH?

In the Arctic, among the ruins of a Soviet research station, there's a 12-kilometre-deep hole. It was welded shut and abandoned shortly after the collapse of the Soviet Union. Decades later, it remains the deepest hole anyone has ever drilled. Locals say it leads to hell.

In the Arctic, there's a 12-km-deep hole. It was welded shut and abandoned shortly after the collapse of the Soviet Union.

It actually leads nearly halfway to the Moho, the boundary between the crust and the mantle. During the Cold War, as the space race heated up, so did rival ambitions to explore below ground. In 1961, a group of geologists from the United States decided to take a 'shortcut' to the mantle by drilling through the ocean floor, where the crust is thinnest (about six kilometres deep). Project Mohole, as it became known, was the brainchild of the American Miscellaneous Society, a drinking club made up of some of the day's leading scientists. 'It sounded so simple and logical at a breakfast meeting on a sunny patio,' member Willard Bascom later wrote. It turned out to be anything but.

This was before proper deep-sea oil and gas exploration. The technology the scientists needed, say, to keep a ship steady while a drill bored deep below had to be improvised, which eventually led to breakthroughs in underwater drilling. The team started drilling off the coast of Mexico and the deep sediment they managed to bring up helped revolutionise geology and the study of ancient oceans.

But the US Congress pulled Project Mohole's funding in 1966. It had spent about half a billion in today's money, and made it only a hundred metres or so into the crust. NASA's space program, then gobbling up the bulk of scientific funding, put humans on the Moon less than three years later.

The Russians got further, drilling down 12.2 kilometres more than a decade later. They were working on land, where the crust is thicker but the job more straightforward. The problem was that their instruments started to melt. 'The heat down there was almost twice as high as they expected,' Tkalcic says. 'It rises 25 degrees for every kilometre you go down.'

German scientists made their own attempt in Bavaria but abandoned the borehole at nine kilometres in 1994 when the temperature hit 265 degrees Celsius. Today it's still open—part tourist attraction, part art installation. (A Dutch artist had a thermally shielded microphone winched down; it picked up a deep rumbling sound scientists apparently couldn't explain.)

Momentum is now building around another push to the Moho, this time by an international exploration program led by Japan. Its science agency has already drilled more than three kilometres into the sea floor, the furthest into the oceanic crust yet, using its flagship vessel *Chikyu*. The ship was designed to drill seven kilometres into the Earth.

Of course, getting to the Moho won't solve every underground mystery. For example, the two blobs around the outer edge of the core, one beneath Africa and the other under the Pacific Ocean, still perplex scientists. 'We could never actually get down there to check what they are,' says Deuss. 'It's all detective work.'

To journey deeper than the Moho, Tkalcic notes, you'd need instruments and probes made of a material that could withstand the same extreme conditions that forge diamonds and melt metal. It's something 'that likely doesn't exist yet', he says. That rules out humans too. 'It won't be like that movie *The Core*. We won't get to ride along.'

HOW DOES EARTH'S MAGNETIC FIELD WORK?

Professor John Tarduno, a geophysicist at the University of Rochester in the United States, often goes hunting for ancient crystals, from South America to Africa. With the right crystal of the right age, he can peer back in time to better understand the Earth's magnetic field. The planet's natural magnetism helps everything from aircraft to turtles navigate and shields the surface from deadly cosmic radiation. When we see auroras twist in brilliant colour in the skies above the northern and southern polar regions, we're witnessing the magnetic field kicking out those harmful solar rays. 'And rocks are like the magnetic strip of a cassette tape, full of tiny magnetic grains,' Tarduno says. 'Some are perfectly formed, ideal for recording the field. We're always searching for them.'

Tarduno and his team found 'perfect' crystals in Quebec, Canada—full of tiny magnetic needles too small to see with the naked eye but still frozen in the direction of the magnetic field's 'north' when they formed 565 million years ago. Back in his magnetically shielded lab, Tarduno discovered something astonishing. 'The crystals showed that the Earth's magnetic field was 10 times weaker back then, as if the field itself was on the point of collapse,' he says.

Yet, when he analysed other crystal records from about 30 million years later, the magnetic field had recovered by 70 per cent—fortunate, for that was when life was proliferating on the planet during the Biological Bang. It also coincides, Tarduno says, with scientists' estimate for when the inner core was formed. The magnetic field itself probably got going much earlier, he tells us, after the giant impact with Theia that birthed the Moon and turned Earth into a molten sea. That calamity may have broken up the layers within our planet, mixing them into the twisting convection needed for magnetism. It's easy for physicists to imagine the magnetic field starting once the molten core was swirling, says Tarduno, 'but magnetic fields decay and fade away too.'

Instead, Earth cooled enough to begin growing a hard inner core. That transformation, from liquid back to solid, creates its own energy, its own heat, as does the shedding of lighter particles within the inner core when the iron crystallises, like salt expelled from freezing water. This energy churns the liquid outer core around it into a boiling sea of convection. 'It's not a river flowing smoothly from one side to another, it's swirling,' Tarduno says. This action generates the magnetic field. The inner core was the engine of this reaction. 'Those conditions could be more of Earth's great luck. If the composition down there had been just a little different, Earth may have met the same fate as Mars.' The magnetic field on Mars died long ago and today its protective atmosphere has largely leaked away.

Deuss says eventually, over billions of years, the core may eat all the way through the liquid core and the magnetic field may die. Other calamities are due to befall the Earth before then—not least of all our sun growing so hot in about a billion years that it cooks the planets around it—but Tarduno agrees that 'as the inner core gets larger, the magnetic field could become more complicated'.

Every hundred thousand years or so on average ('though very much at random,' Tarduno notes), the field switches poles. North becomes south. It often begins with an anomaly, an area of the field pointing in the wrong direction that grows over thousands of years until it takes over, triggering a rapid reversal. One such anomaly has been detected in the southern hemisphere, over the Atlantic Ocean, leading many scientists to predict a coming reversal. The magnetic field can drop in intensity during these periods before jumping back up, which could spell trouble. But Tarduno says that, while the field has been waning for the past 160-odd years, the Earth's own atmosphere still shields us well from cosmic rays. 'There have been hundreds of reversals in the last 100 million years.' Still, today a reversal could send technology haywire.

Tarduno believes the cause of the reversals lies in the core-mantle boundary—with the two strange blobs. The South Atlantic anomaly lines up with the African blob, which may be pushing on the core below it, changing the heat flow, the convection. The boundary between rock and liquid there isn't a perfect sphere inside the Earth. It has topography, says Tarduno, like upside-down mountains.

It's also the most dramatic boundary within the Earth, Tkalcic says, 'more, even, than going from atmosphere to solid rock'. Earth isn't perfectly symmetrical either, he adds. 'You have things like the blobs [affecting] how its mass, its gravity, is centred.' That causes the magnetic poles to wander a little. And it tangles the invisible lines of the magnetic field. When they become too tangled, he says, just as a build-up of pressure between two tectonic plates triggers an earthquake, 'you have a release in the system, a sudden swap'.

WHAT OTHER BIG MYSTERIES REMAIN?

Unlocking the secrets of our planet's interior could help us understand life on the surface and how it got started. (Living microbes have already been discovered deep in the heat of the crust, for example.) Meanwhile, many of the rare metals and other resources left in the Earth are lower down and many countries, including China, hope to map more of this subterranean world.

Unlocking the secrets of our planet's interior could help us understand life on the surface and how it got started.

Deuss wants to see if a link exists between changes within the inner core and those in the magnetic field. Suspended in liquid, the inner core spins independently of the Earth, at times faster and at other times slower. Those fluctuations seem to line up with a small wobble the Earth experiences in its rotation every six years. And the inner core appears to be growing faster on one side than the other.

Scientists are trying to simulate the core's punishing conditions in their laboratories, using diamonds and shock-waves created by powerful gas guns. Already, Japanese scientists have helped confirm theories that the inner core is a forest of crystal iron. 'You basically take two diamonds and push them together, tighter and tighter,' explains Deuss. 'Then you add lasers to make it hotter too. We'd been able to produce pressure and temperature but never at the same time before [those scientists] cracked it. The diamonds would just break.'

But she thinks more secrets could be revealed at an even smaller scale—using tiny particles born from radioactive decay called neutrinos. Most of the time, these tiny 'ghosts' wing through space at a constant speed. But every now and then they hit an atom of matter and when they do, they can tell us something about that atom, Deuss says. Finding a neutrino that has touched even one atom of the inner core could be a big jump forward for the field.

'Because now I've been looking at the [seismological] data and there's just too much noise, we've seen as deep as we can, I think,' Deuss says. 'We need next-generation detectors, like for these neutrinos; a few in the right places.' Her ideal spot? Above each of the mystery blobs, waiting to catch a stray neutrino. Tarduno suspects these 'oddities' are not the remnants of a lost planet but slabs of the first super-continent Pangea. If proven, this theory would complete a kind of circuit for our living Earth, he says, as things at the surface sink to the mantle, pressing on the core and, in turn, changing conditions at the surface.

But whatever mysteries lurk below the Earth's surface, Tkalcic thinks Jules Verne was right about one thing: 'It's still a whole other world down there.'

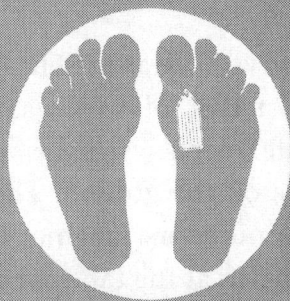

14

WHAT HAPPENS IN AN AUTOPSY?

Sometimes when people die, questions
linger over how and why. An autopsy
can provide answers. How are
they performed?

Jackson Graham

No one saw Tyson Manders fall. One minute, the 26-year-old father of two was motorbike riding with his wife's nephews at their farm. The next, he was lying motionless on the ground. The family had been celebrating a birthday outdoors, laughing over lunch. Tyson's wife, Georgie, was sitting at the table when her nephew ran up calling for help. The moment she saw Tyson, she knew he was dead. 'I just had this feeling,' Georgie says. 'He was still and looked peaceful.'

Time slowed for Georgie as her father and sister tried to resuscitate Tyson. Paramedics confirmed her worst fears—the husband she adored was gone. Tyson and her nephews had been riding carefully so a freak fall and an awkward landing seemed, at first, the most likely explanation for his death. Tyson's body was taken to Sydney for examination.

Days later, Georgie received a call from the New South Wales state mortuary to say there were no signs of any injuries to suggest a bike accident caused Tyson's death. Would she allow an autopsy? 'At first, I was firmly thinking no,' she says. 'I didn't want Tyson's body being tampered with, I just wanted him home. But I kept coming back around to being haunted with the question, "Why Tyson?" There has to be a reason behind this. None of it made sense.'

Whether or not someone we love or know might one day need an autopsy isn't something we tend to think about. Most deaths, after all, don't require a formal investigation. Yet in some cases, it's a legal requirement. Forensic pathologists examined about 12,000 bodies in Australia in 2023, out of a total 182,000 deaths recorded that year.

What happens in an autopsy? Who needs one? And how can investigating a death help other people?

WHAT'S AN AUTOPSY?

In a mortuary in Melbourne, a technician gently wipes fluid from the mouth of a deceased man. Meanwhile, forensic pathologist Noel Woodford peers at the face, speaking into a voice recorder. The body is here because the man died unexpectedly. The coroner has authorised the autopsy after considering both the circumstances of the man's death and any family objections. In this case, his death was unexplained and his family had no objections to the autopsy, says Woodford, who leads the Victorian Institute of Forensic Medicine.

During an autopsy, from the Greek 'to see for oneself', pathologists try to answer critical questions: How did this person die? What happened in the body leading up to death? And, sometimes, what were the circumstances surrounding the death? It can involve inspecting injuries on the surface of the body; opening the body to look at a particular area; or, in full autopsies, examining all organs in the chest and abdominal cavity, and sometimes the brain. TV mysteries often show forensic pathologists puzzling over murders, but in real life most of their work involves natural deaths.

TV mysteries often show forensic pathologists puzzling over murders, but in real life most of their work involves natural deaths.

At the Melbourne autopsy, the man's body faces upwards under bright theatre lights. The skin is pale but the back and the undersides of the arms and legs are crimson because of a process called lividity, in which blood pools at the lowest point of the body after circulation stops. Woodford looks closely at the neck, torso and limbs, noting that there are no injuries. He examines the skin for signs of disease or scarring.

The technician makes an incision with a scalpel on each side of the chest then cuts a single line down to the waist, forming a Y. He pulls back the skin, exposing a layer of fat and muscle, then uses clippers to remove a section of the rib cage. Most of the organs are now exposed. The bowel comes out first. 'By taking [the bowel] off, we get a better look at the structures on the abdominal wall inside,' Woodford says.

When families ask what happens in an autopsy, Jodie Leditschke, the coronial admissions and enquiries manager, describes it as a surgical procedure. 'But then we often say to them, "What do you want to know?", because some families want to know exactly what's happened,' she tells us. 'We have often described to families exactly where the incisions are.'

One technique, developed by 19th-century Austrian physician Karl Rokitansky, involves removing all the organs, from the tongue to the rectum, as an intact block. It's done deftly and with great care. 'We are always making sure we don't hurt the deceased, even though we know, in theory, that's not possible,' Leditschke says. The organs are inspected for clues that can be found in their relationship to one another—the path of an infection, for example, might be visible from one organ to another.

The sight of this block of tissue—the workings inside all of us—might be confronting to the untrained eye. To Woodford, it can reveal what he can't see on a scan: the full extent of old and new plaque in arteries, say, or the grainy texture of kidneys scarred by high blood pressure. An autopsy, to him, is 'a stepwise, regimented approach to looking at tissues'. His observations can help the person's family understand what happened to their loved one. 'If we can do our bit, which is providing answers—not just to the coroner or the police, but to them—we feel like we've done a good job.'

WHO NEEDS AN AUTOPSY?

One of the first recorded forensic postmortems was conducted on Julius Caesar after his assassination in Rome in 44 BC. The physician, Antistius, concluded that, of the emperor's 23 stab wounds, one to the breast had been fatal, although exactly who delivered it remained unknown. In cultures where prompt burials are customary, such as Jewish and Muslim communities, autopsies were traditionally forbidden. By the Renaissance, dissections of human bodies were a regular part of European medical education; the Amsterdam Guild of Surgeons held a public dissection annually. The 1632 event, featuring the body of a man hanged for stealing a winter coat, is immortalised in Rembrandt's *The Anatomy Lesson of Dr Nicolaes Tulp*.

Today in Australia, a death is reported to a coroner, usually by police, when it was unexpected, unnatural or violent. A report also follows someone dying as an unplanned outcome of a medical procedure or while in custody or state care. The coroner takes custody of the body to confirm the identity of the person and ascertain what caused their death and, if possible, the circumstances surrounding it. 'We are effectively taking the body into our care so investigations can be made to provide answers to those questions,' says Victoria's State Coroner, John Cain.

'We are always making sure we don't hurt the deceased, even though we know, in theory, that's not possible.'

Most natural deaths don't require investigation if a doctor is satisfied they can record a cause. Queen Elizabeth II's death certificate, for example, simply stated 'old age'. A natural death is examined if the person's medical history can't explain how they died, says Linda Iles, who heads forensic pathology services at the Victorian Institute of Forensic Medicine. 'Others may have a known chronic disease but a GP may be unwilling, for a number of reasons, to issue a death certificate,' she says. Accidental deaths, including falls, car accidents and workplace incidents, are also common cases in the mortuary.

At the institute, the coroner meets with pathologists and staff overseeing admissions and enquiries to 'triage' new cases. All suspicious deaths, and some that are not suspicious but are confounding, receive an autopsy. For others, a CT scan, a toxicology report, an external examination and information from police and doctors might be sufficient to determine the cause of death.

The growing use of these non-invasive examinations has reduced the number of autopsies. In the early 2000s, for example, nearly every body that arrived at Victoria's state mortuary would undergo an autopsy, but now the autopsy rate is between 35 and 45 per cent—roughly 3000 a year. Initial tests and scans can also rule out causes of death. Queensland man Donald Morrison died after removing a snake from a friend's leg in 2023, but findings showed there was no venom in his system. A coroner's investigation found he'd died of natural causes.

When a non-suspicious case does require an autopsy, the coroner takes into account the wishes of the dead person's loved ones before ordering the procedure. 'There are those that are very anxious to know what was the ultimate cause of death. And then there're those that say, "Look, the person's deceased, we're very sad about it; we just want to get on with the funeral and don't need to know the precise details",' Cain says. 'Unless there's some other strong reason to have an autopsy, then we'll respect the family's views—and it's generally around religious or cultural views.' Customs that call for burial soon after death are still a common reason for an objection.

DOES AN AUTOPSY ALWAYS PROVIDE AN ANSWER?

In the mortuary, Woodford weighs each organ—the heavier it is, the greater the possibility of disease—and examines each for signs of disease, including cancer. He makes tiny incisions on the coronary arteries to check for clots or blockages. As he examines each organ, he takes a nick of tissue that will be scrutinised later under a microscope. Samples of blood and urine are also collected for toxicology tests. Meanwhile, the technician photographs the organs so another pathologist can review the case if the family or authorities request it.

The procedure takes just over an hour. (A suspicious death takes longer as it involves a more detailed examination and more photography.) Then the technician places cotton wool in the chest and neck cavity, Woodford and the technician return the organs to the body, replace the chest plate and sew the skin back together. 'It's important for us to ensure the body is returned to a whole state as much as we can,' Leditschke says. 'If we retain organs, we contact the families to let them know.'

Afterwards, just a thin line of neat stitches is visible. 'A family, even though they are aware an autopsy has taken place, actually shouldn't be able to tell, if he was wearing a singlet or an open-[collared] shirt,' Leditschke says. The technician washes the body with running water, lifting each arm carefully and dries the skin with towels. Then he places a shroud over the body and wheels it from the room. It will likely now be released, with a provisional cause of death if the coroner allows. In Victoria, the body is generally held in the mortuary for between four and seven days before the coroner releases it to a funeral home.

The body of Woodford's own father came to the mortuary after he died of a stroke. It was the 1990s, and Woodford, who had been a trainee at the institute, was by then working elsewhere. The experience showed him what the process was like from the outside, as a family member. 'I was really amazed at how respectful everybody was, and sympathetic.' He also appreciated how important the death certificate was for managing his father's finances. In cases where a coroner has yet to establish a cause of death, the Registry of Births, Deaths and Marriages can issue an interim death certificate, although it might not be accepted in all financial situations.

Pathologists often find a single underlying cause of death, such as a clot in an artery. More puzzling cases can involve a confluence of events—death certificates typically ask for a hierarchy of events but can allow for several causes to contribute equally. 'As we get older, we have more things going on,' Iles says. 'Or, in younger people, say, you've got one disease process going on but you've got some drug use going on as well. Or you find yourself in a physical circumstance that is adverse. It can become quite complex.'

If an autopsy and other scans and tests don't provide a conclusive answer, the case can be referred to further specialists. Genetic testing might reveal heart rhythm disorders that can't be detected under a microscope, such as Long QT syndrome, Brugada syndrome or the tachycardia known as CPVT. In about 5 per cent of cases, Woodford says, the cause of death remains 'unascertained', although an autopsy can at least exclude causes such as trauma or infection.

Once an autopsy report is complete, the coroner either finalises the case or considers it as part of a wider investigation. 'It can open up other lines of inquiry that go to the circumstances in which somebody died,' Cain says. The coroner can hold an inquest—at which a court hears evidence—or make written findings that are published based on public interest. In other cases, the autopsy report becomes part of a criminal investigation and prosecution.

HOW ACCURATE ARE AUTOPSIES ON TV CRIME SHOWS?

Sydney forensic pathologist Johan Duflou felt his 'spidey senses tingling' while examining the body of an elderly woman in 1989. She had injuries to the back of her head that didn't appear likely to have happened in a typical fall. The police attended the autopsy—a wallet had gone missing—and Duflou recalls that police raised the theory that the death was the result of a 'mugging gone wrong'.

Still, the pattern of the injuries bothered Duflou. He told police he suspected it was a homicide. The woman turned out to be a victim of Sydney's so-called 'granny killer', John Wayne Glover, who was convicted in 1990 of murdering six women. Glover hit several of his victims with a hammer. 'Obviously, we're not police officers,' Duflou says. 'But we investigate deaths. Every so often, we come up with something which is of major concern.'

Crime-show portrayals of forensic analysts as world-weary oddballs don't sit well with real-life practitioners. 'It is not only misconceived, it makes us

'Of the forensic pathologists we have on staff, our youngest are young women. We don't have any of the crusty old professor type at all.'

very much annoyed to see us portrayed that way,' Duflou says. 'We're all different. Some of us have more foibles than others.' Says Iles: 'When you look at the forensic pathologists we have on staff here, our youngest are young women. We don't have any of the crusty old professor type at all.'

As on TV, pathologists do occasionally attend police scenes, but only if there's a medical or investigative question they can help address, such as whether a death might be suspicious. Cain will be notified of deaths such as a police shooting or someone dying in custody so he can attend the scene if he chooses. 'You just get a much better idea of what's happened,' he says.

In real-life, autopsies for suspected homicides will be fast-tracked, with pathologists and police working together closely: without context from police and the scene, whether someone was assaulted, had an accident or inflicted a wound on themselves can be difficult to distinguish. 'It's whether the injuries make sense in the context it's been put to you,' Iles says. The devil can be in the detail. If someone has injuries on a single part of their face, they might have fallen, whereas a fight or attack can leave scattered injuries. 'It's not always the actual fatal wound itself but potentially smaller injuries that give you information,' says Iles.

The autopsy will also consider whether the person's health played a role in their death. 'It's not exactly rocket science that they've been shot or stabbed,' says Iles, 'but it's the [surrounding information] that the autopsy is for—not just for the prosecution but for the defence.' The hands, fingernails or injuries themselves may carry trace evidence that helps police identify suspects or objects of interest. Pathologists take swabs but are rarely privy to the results. 'We're not necessarily about who did it,' Iles says.

HOW DO THE EXPERTS DEAL WITH SEEING DEATH UP CLOSE?

It can be a challenging field, but the pathologists we interviewed say their work isn't different from that of emergency or medical workers who confront similar scenes. 'When it comes to seeing confronting and, you know, potentially quite gruesome things, I have a job to do and it's to solve problems,' Iles says. Roger Byard, an emeritus professor at the University of Adelaide's School of Biomedicine, puts it this way: 'The way that pathologists survive is because we treat it as a scientific exercise.'

Some cases can affect these professionals, though. Byard recalls a young woman who he discovered had been buried alive. Her ex-boyfriend subsequently pleaded guilty to murder and was sentenced to jail for 22 years without parole. Byard received a forensic science award for his work on the case, but its horrible circumstances have lingered with him. 'I did come up with the right diagnosis. And yes, he did plead guilty. But it's a case that I find very difficult to find anything positive about.'

Leditschke's first day as a technician at the Melbourne mortuary in 1988 went smoothly—'I was so excited'—but the next day she passed out. 'It was sights, sounds, realising that I'm standing there among so many deceased persons,' she recalls. The most affecting cases, she says, include people who die without family and youngsters who die at similar ages to children in the lives of mortuary workers. In some of the hardest cases,

> The most affecting cases include people who die without family and youngsters who die at similar ages to children . . . of mortuary workers.

technicians make plaster casts of the feet of infants for the families. 'That's just something the mortuary staff took up,' Leditschke tells us.

After decades of working with technicians, and now managing a team that speaks to grieving families, Leditschke sees the job from both sides: 'To ensure you respect the body . . . you can't completely [say] this is just anatomy,' she says. 'What you're dealing with is a recent death where, right at that time, there is a family grieving.'

Byard has done more than 6000 autopsies but, even with all his experience, kissing the forehead of his deceased father still shocked him. 'He was so cold and waxy,' he says. 'Death can be confronting even to somebody who deals with it every day. I was looking at him in the funeral home, thinking, *It has to be him*. But he looked different. Death is a difficult process.'

HOW CAN UNDERSTANDING DEATH HELP THE LIVING?

On the farm where Tyson Manders died, in Yetholme, near Bathurst, a garden grows with a sign: 'Daddy's spot'. This is where Georgie and their two children go to remember him. Georgie's family has kept bees there for four generations, and at the centre of the memorial is a beehive box in which the family deposit letters and mementoes such as carved pumpkins on Halloween. 'He was a very hard-working, fun-loving, seemingly healthy, strong young man,' Georgie says.

The autopsy revealed that a heart condition called hyper-trophic cardiomyopathy caused Tyson's death. Counsellors told Georgie it was probably instant. 'I found a lot of peace of mind knowing that,' she says. 'We all held a lot of guilt thinking at first that it was a bike accident. We were doing CPR, we were doing everything we could. Did we overlook something? There was nothing we could have done . . . He wouldn't have experienced any pain.'

Often a deceased person's loved ones will ask mortuary workers whether there is anything they could have done to save the person. 'More often than not, there isn't,' says Natalie Morgan, a family liaison nurse at the Victorian Institute of Forensic Medicine. 'We always tell the truth, but you do try to emphasise reassurance wherever possible.' Sometimes loved ones want to know why the body appeared a certain way; if they saw the body with blood coming from the nose or mouth they might believe the person was injured. Morgan explains this is often due to a sudden fluid build-up after the heart stops. 'It can just be part of the normal process of dying.'

People often ask about the time of death. Iles says crime shows have perpetuated the myth that this is something pathologists can accurately determine. 'Occasionally, people have pacemakers and bits and pieces and we can interrogate those,' she explains. 'Circumstantial information is often a lot more reliable than the rate of decomposition.'

People often ask about the time of death. Crime shows have perpetuated the myth that this is something pathologists can accurately determine.

In some cases, an autopsy flags the need for genetic screening of family members, including for cancers, neurological conditions and heart disease. About one in 500 people carry a genetic variation that could lead to a heart condition. Georgie now knows Tyson carried a gene for hypertrophic cardiomyopathy that has a 50 per cent chance of being passed to their children, Elaina and Aaron.

Tyson's autopsy gave Georgie a sense of control: 'I think knowledge is power'.

The children now receive regular heart checks and will undergo genetic testing. If they carry the gene, medication can help manage the condition. 'Tyson has left behind a legacy,' Georgie says. 'My husband's life and the father of our children will protect our future generations just with that knowledge of the cause behind his death.'

15

HOW DO YOU BUILD A GREAT SANDCASTLE?

Many minds, from Buddhists to Beyoncé, have dug into the meaning of sandcastles. But what's the best way to actually make one you can be proud of?

Angus Holland

'Sand ain't just sand,' says Peter Redmond.

St Kilda Beach, near the pier?

'Trash. There's about 100 millimetres you can use, but the rest is just really grainy, gluggy, crappy.'

Bondi?

'Beautiful sand to work with because it's had a billion people walking over it constantly. Manly is the same.'

And the sand in Western Australia?

'Totally different—what you would call a sharp sand.'

He means that if you study a sprinkle of Western Australian grains under the microscope, they look like miniature crystals—not rounded-off kitty-litter pellets but all edges, corners and facets. Which, in the world of professional sand sculptors, is a good thing, since they kind of lock together like Lego.

That said, competition-winning sand sculptors—the Formula One drivers of sandcastle building—such as Redmond don't use plain old beach sand at all. Instead, they order up batches of similar but much more consistent sand, which enables them to craft elaborate structures that would otherwise collapse.

Yet their techniques and know-how can still help us rank amateurs as we while away an afternoon on our seaside holiday. How can we all build a tip-top sandcastle? Why would we want to? And what happens when the tide washes in?

WHAT CAN SANDCASTLES TEACH US ABOUT HUMAN NATURE?

The transience of sandcastles—painstakingly built only for the tide, an errant dog or a toddler to sweep them away—has long been a go-to metaphor for the fragility of existence.

German philosopher Friedrich Nietzsche wrote of a primordial united force or god-type creature who, like a child, 'sets down stones here, there, and the next place, and who builds up piles of sand only to knock them down again'. In this way, observes Iranian philosopher Hamidreza Mahboobi Arani, Nietzsche

The transience of sandcastles has long been a metaphor for the fragility of existence.

suggests 'the entire universe, including ourselves, is nothing more than a momentary configuration of shapes in the sand'.

In a Buddhist children's story, *The Secret of the Sand Castles*, children at the beach start arguing about whose sandcastle is best and—yes—end up trampling the lot. A 'magical' wise man arrives to teach them that the sandcastles did not matter nearly as much as creating a harmonious world with each other.

A castle etched on a grain of sand. *Vik Muniz, Marcelo Coelho*

The late US psychoanalyst Stephen Mitchell once used sandcastles as an analogy for a relationship, which he called a 'sandcastle for two'—requiring constant maintenance, an awareness of life's impermanence and an understanding that the other person, as with a sandcastle, ultimately 'lies outside one's control'. Drawing an even longer bow in his book *Darwin's Sandcastle,* author Gordon Wilson argues his case against the theory of evolution: 'It's about time this philosophy is seen for what it is: a sandcastle on the beach, in the face of the rising tide.'

Musicians, too, are drawn to the analogy. In his 1964 song 'Castles in the Sand', Stevie Wonder warned listeners that romances generally lasted about as long as you know what. Beyoncé echoed the theme even more gloomily in 'Sandcastles' from 2016, singing of how a couple's relationship, as with the sandcastles they'd built together, ultimately ended in tears.

Metaphors aside, many people who make sandcastles and related sand art (which can take a multitude of forms) seem to enjoy their structures' ephemeral nature: the opportunity to build and enjoy then destroy and start anew, no two constructions exactly alike.

ENOUGH OF THE METAPHYSICAL, WHAT ABOUT THE PHYSICAL?

Scientists and engineers have taken as much interest in the humble sandcastle as philosophers, not only to find the perfect construction technique but also to understand how the behaviour of sand grains can have wider applications.

Sand is one of our most essential resources. 'Without sand, we couldn't have contemporary civilisation,' writes Vince Beiser in *The World in a Grain.* It's the raw material for, among other things, concrete, glass, the silicon chips inside our phones and computers and 'even the glue that makes your sticky notes stick', he writes.

Sand is one of our most essential resources. Without sand, we couldn't have contemporary civilisation.

Sand, for the record, is defined as (deep breath here) a naturally occurring granular material composed of fine weathered mineral particles. One common definition states that it's 'sand' if the particles are between 0.0625 millimetres and 2 millimetres in size. Larger particles are gravel; smaller ones are clays or silt.

The most common form of sand is predominantly quartz, although sands can be made up of bits of coral, volcanic detritus, natural glass and the ground-down shells of long-dead sea creatures, known as calcium carbonate. In the Caribbean and the Hawaiian islands, most of the white sand beaches were created by parrotfish eating reef coral and excreting it as sand (there is black volcanic sand too).

As previously suggested, the shape of each particle can vary dramatically. Desert sand is windblown and rounded, which makes the grains less able to stick together; marine sand, particularly river or glacial sand, is typically sharper and can contain fine silt (which helps the grains lock together) and is prized by sandcastle aficionados.

Sand is unusual in that it can be both super-firm—like the stuff we walk on at the beach—or behave like a fluid, such as when it flows through an hourglass, says Dr Francois Guillard, a particles and grains expert at the University of Sydney. Things get really interesting, though, when you mix in a little water and the sand particles, water and air start to cohere.

On a microscopic level, the liquid's capillary effect—a combination of surface tension and adhesion—bridges the gaps between the grains of sand, loosely bonding them. The rules of 'capillary condensation' were first laid down by British mathematician and engineer William Thomson, better known today as Lord Kelvin, in 1871. They were updated in 2020 by a team of researchers at the University of Manchester, led by Nobel Prize winner Andre Geim, who stated: 'Such important properties as friction, adhesion, stiction, lubrication and corrosion are strongly affected, if not governed, by capillary condensation. This phenomenon is important in many technological processes used by microelectronics, pharmaceuticals, food and other industries—*and even sandcastles could not be built by children if not for capillary condensation.*' Italics are our own.

Add too much liquid, though, and the capillary bonds start to fail, 'and you just have water everywhere', says Guillard.

This is useful to know, and not just when you're building a sandcastle. How particles behave affects many industries, from pharmaceuticals to construction. 'If I have a bulldozer that is pushing a pile of sand,' Guillard says, 'what kind of force is that going to require to move the material?' He points out that the world's tallest building, the Burj Khalifa in Dubai (at nearly 830 metres), has Australian sand in its concrete—the local stuff, made of grains rounded for millennia by the wind, was ruled out. 'Obviously, there's a lot of sand in the desert nearby, but it's not really suitable.'

The world's tallest building, the Burj Khalifa in Dubai, has Australian sand in its concrete.

When you build your castle, getting the ratio of sand to water right is critical. That said, how much water you use is still contested. Some experts prefer the 'just enough' theory, which takes into account the type of sand in a particular location and—another wildcard—the amount of salt in the water, which may have its own mildly adhesive properties.

In the lab, Dutch physicist Daniel Bonn found he could build the highest towers with just one part of water to 99 parts of sand. Bonn was also part of a team that performed experiments to demonstrate how the ancient Egyptians might have made it easier to move the massive stone blocks used to build pyramids—by adding a little water to sand underfoot to make it more solid. This might explain a wall painting found in the tomb of Djehutihotep that shows a worker pouring water onto the sand in front of a sled carrying an enormous statue. (For years, Egyptologists assumed it was some kind of purification ritual.)

In 2008, researchers at the Max Planck Institute for Dynamics and Self-Organisation in Gottingen, Germany, used X-ray microtomography (which creates 3D models of microscopic objects) to show that the precise amount of water between grains didn't matter as much as previously thought. From less than 1 per cent to more than 10 per cent water, they found, the sand-water-air mass remained just as stiff. 'These properties are not only significant for the construction of sandcastles,' said Professor Stephan Herminghaus, the leader of the Max Planck Institute team. 'Wet granulates are relevant in very many fields, and now we have a better understanding of their mechanical properties.'

Matthew Bennett, a professor of environmental and geographical sciences at Bournemouth University in England, suggests one bucket of water to eight buckets of dry sand is probably the most effective ratio for adhesion—it's typically the sand–water mix you find at the high tide line. Bennett knows this because in 2004 he was commissioned by a travel company to find Britain's best spot for building sandcastles, unexpectedly propelling him to media stardom and making him known henceforth as 'Professor Sandcastle'.

Given the demands of his day job—he is a specialist in prehistoric footprints and has published widely in the fields of glaciology, sedimentology and geomorphology—Bennett initially found the sudden burst of attention slightly galling. 'If the public want fluff, let them have it, better something than nothing, right?' he wrote on his blog. Nearly 20 years later, he tells us that he appreciates the episode's upside. 'Sandcastles is not a bad vehicle to talk to the public about science. What works is the emotional connection, not how sophisticated the research is or where it is published. That is the lesson in the sandcastle saga: most folks have played in the sand, it is something folks can connect to.'

SO, HOMEWORK DONE, WHERE DO I START DIGGING?

It's slightly counter-intuitive, but you don't make top-notch sandcastles by building them from the ground up. Rather, you go from the top down. Think of your castle as a sculpture made of marble—Michelangelo's *David*, perhaps—where you start with a single block of stone and chip away until your masterpiece is revealed.

You don't make top-notch sandcastles by building them from the ground up. You go from the top down.

There is no great secret to it, says Peter Redmond, who has constructed, among other creations, a singing kookaburra and a snowman for Queensland's Townsville Council. 'When you're sculpting, you either add sand or you take sand away,' he says. 'There's no insider knowledge or . . . some sort of skill that nobody else can do.'

So start by building a big mound, ideally of sand that is already moist. Then add several bucketloads of water.

Experts such as Redmond do have a slight advantage over the rest of us: they typically truck in their preferred kind of sand, many tonnes of it, then build it into a mound in layers, like a wedding cake, using a timber framework to stop it collapsing. Adding water, they pound it down, sometimes with a tamping tool, sometimes by hand, until it's incredibly firm.

That's how you make the sand strong enough to go really big, like the world's largest sandcastle, a 21.16-metre-high, 4860-tonne monolith created by Dutch sculptor Wilfred Stijger and a team of 30 in the Danish seaside town of Blokhus in 2021. (The world's tiniest sandcastles were microscopically etched onto individual grains of sand in 2014 in a collaboration between artist Vik Muniz and Marcelo Coelho, from Massachusetts Institute of Technology, using a focused ion beam and a scanning electron microscope.) One of the world's most sought-after sandcastle builders, American Calvin Seibert, uses whatever beach sand is at hand to create his extraordinary brutalist structures right by the shore, where they can indeed get washed away. He makes a big pile of sand, then sprinkles it with water he gathers from the sea until it reaches Goldilocks consistency: not too wet, not too dry (both states lead to collapses).

Then it's time to get to work.

You will need a bucket and a spade. Compacting the sand as you create your pile will help it stick together, so squish it or stamp it as you mix in the water. One technique is to pour water into holes that you have poked into the sand, then stand on the pile until it feels solid. Seibert can use over 700 litres of water during a build.

WHERE DO YOU START SCULPTING?

The next phase is to shape the basic form: as in, chip away at the chunks of 'marble' you know you definitely won't need, from the top. You should have at least a rough idea of the overall size and shape of your sculpture, says Redmond, so it doesn't end up out of proportion. 'We had a young guy do some work with us a few years ago—really talented, he does modelling for the toy industry,' says Redmond. 'And he was doing a gorilla for us. We kept telling him he was going to run out of space and maybe he should shrink it down a little bit. But he didn't listen and when he got down near the bottom, he didn't have room for legs.' He solved the problem by 'burying' the beast.

HOW DO YOU MAKE ALL THE FIDDLY BITS?

Once your basic outline takes shape—whether it's a traditional castle, a gorilla or a gorilla-dragon sand sculpture—it's time to add the fine details. Though if you're working on a very big construction, you might want to work on the top part first, says Meg Murray, a champion Tasmanian sand sculptor who has exhibited internationally. 'For example, with a big elephant, you just block out a rough shape with no detail, the top half first. You then pick up smaller tools and start defining detail. Then start blocking out again, removing and blocking and then detailing as you work your way down to the bottom.'

As you go, you might want to add additional touches, such as an arch between two towers. Using wet sand, slop a little on either side of the span, slowly building out from the towers towards the centre. You can draw brickwork patterns using a plastic knife, a paint scraper or a small trowel. A paintbrush or drinking straw can be used to shift unwanted grains. Seibert uses scraps of plexiglass.

When you're done, stand back and enjoy—because it won't last for long.

Sometimes professionals spray their finished piece with a glue solution to protect it from the rain. A US inventor, Daniel Perlman, even patented a non-toxic, environmentally friendly additive (made from pre-gelatinised starches, chemically modified starches and chemically modified celluloses) which, mixed with sand and water, promised to make a sandcastle last indefinitely, or until it was deliberately doused with water. 'A company that found other patents of mine contacted me and asked if I could find a way for children playing at the beach to build stronger and taller sandcastles and other sand shapes that could be dried but eventually collapsed with water if and when desired,' Perlman tells us. It never went into commercial production. But, he says: 'I'd be happy to be introduced to a new company, perhaps in Australia, where sand and beach activities seem to be taken more "seriously".'

Not everybody is that fussed about longevity. 'It's this moment that we share with others, and I love that aspect,' says Meg Murray. 'It's not a piece of art that people put a price on.' She never makes the same piece twice, if she can help it. Indeed, Calvin Seibert once wrote: 'You can't be doing what I'm doing without getting used to loss. I could turn around and walk 30 feet and my work could be gone, destroyed by a wave, seagulls, a child. But sometimes I come back days later and the sandcastles are untouched.'

16

WHEN WILL WE REACH 'PEAK' POPULATION?

The world's population will stop growing—but when? And where will all the people be in 2100?

Matt Wade and Angus Holland

It's 2050 and the world is home to nearly 10 billion people. India leads the way with a head count of 1.69 billion, more than China and the United States combined. But Africa has become the world's population juggernaut, with a quarter of the world's people and the largest workforce of any continent. Many countries in Europe and Asia are ageing and shrinking. Australia has managed to hold its own, barely shifting in the rankings as other nations move up and down.

This forecast might seem abstract, but population counts: it shapes cultures, affects economies and influences political clout. That's why it's a perennial source of public anxiety, whether it's growing or shrinking. There have long been fears about too many people on the planet. Unchecked population growth would lead to poverty, misery and war, according to 18th-century British economist Thomas Malthus. 'Malthusian' pessimism has emerged once again in the era of climate change.

In 2023, Japan's prime minister warned that his country was 'on the brink of being unable to maintain social functions' due to its super-low birthrate.

Fears about not enough people are perennial too, now keenly felt in an increasing number of countries where declining birthrates have provoked warnings of a 'population winter'. In 2023, Japan's Prime Minister Fumio Kishida warned that his country was 'on the brink of being unable to maintain social functions' due to its super-low birthrate. It is one of many nations in Asia and Europe adopting policies that encourage women to have more babies. In South Korea, private companies are offering workers cash to have children and free cars to employees with big families. Even in Australia, where the situation is less acute, 'it would be better if birthrates were higher', Treasurer Jim Chalmers said in 2024.

In Britain, where birthrates have dropped to their lowest level in two decades, demographer Paul Morland controversially called for a tax on people who *don't* have offspring. 'This may seem unfair on those who can't or won't have children,' he wrote in *The Sunday Times*. 'But it recognises that we all rely on there being a next generation and that everyone should contribute to the cost of creating that generation.'

So how will the world's population change this century? Where will all the people be? And how will these changes affect Australia?

WHAT'S SHAPING THE WORLD'S POPULATION?

Every two years the Population Division of the United Nations updates its projections. The division's first report, in 1951, put the world's population at 2.4 billion.

Since the mid-1970s the world has been adding an extra billion people roughly every 12 years. The head count reached 8.12 billion in 2024. But if you assumed the world's population was on a never-ending upward spiral, think again. The rate of population growth peaked at over 2 per cent a year in the early 1960s; in 2020, it dipped under 1 per cent for the first time since 1950; and it is expected to ease to 0.5 per cent by 2050. The number of people in the world will peak at 10.3 billion in the mid-2080s, according to the UN. After that, growth will sink into the negative.

If you assumed the world's population was on a never-ending upward spiral, think again.

The trend is like a slowing vehicle, says Associate Professor Udoy Saikia.

'I tell my students that, "Yes, we are still driving the car forward but as we move forward, the speed will be less and less and less",' says Saikia, a population expert at Flinders University.

Other forecasts differ, although mostly over timing rather than the overall trend. Modelling by the Institute for Health Metrics and Evaluation (IHME) at the University of Washington in Seattle, for example, forecasts that the world's population will top out at 9.7 billion in 2064 and then drop. According to the IHME, the world's population will be 8.8 billion in 2100, which is 1.4 billion fewer than the UN's central projection. Saikia says significant declines in fertility rates recently, especially in large nations such as China, India and Bangladesh, mean that the world's population might peak earlier than the UN has predicted. After the peak, more people will be dying than are being born.

A fertility rate measures the average number of children a woman can be expected to have during her lifetime. Since 1950 the global rate has more than halved from around five to 2.3. By 2050, it is projected to drop to 2.1. As it happens, a fertility rate of 2.1 is known as the 'replacement level', the point at which births and deaths are roughly in balance, creating a stable population (excluding migration). After 2050 the fertility rate is expected to drop further.

Most people in the world, including in its 15 biggest economies, live in nations with a fertility rate below the replacement level.

Already, the *share* of nations with very low fertility rates has jumped. Most people in the world, not least in its 15 biggest economies (measured in US dollars), live in nations with a fertility rate below the 2.1 replacement level, including China and India. Australia had a rate of about three children per woman in 1970. By 2022 it had fallen to 1.63. Fertility rates have declined even in places associated with rapid population growth, such as Asian megacities. 'Take a city like Kolkata in India, for example,' says Saikia. 'There the fertility rate is now much lower even than in Australia.'

Some low-income countries, especially in Africa, still have relatively high fertility rates. Niger topped the list in 2019 with 6.9 followed by Somalia (6.5) and Chad (6.4). But even in those nations, rates are trending lower. In Botswana, the fertility rate has dropped to around 2.8 children per woman. In Namibia, it's 3.3; in Zimbabwe 3.5, according to the World Bank. All three countries had a fertility rate hovering around seven in 1970.

Here's how the trend looks, with births projected to dip below deaths.

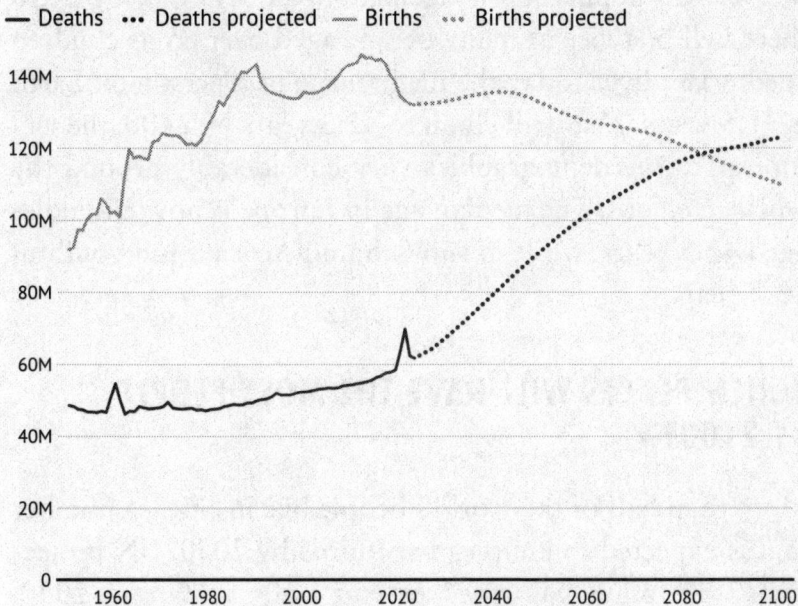

World births and deaths.
Matthew Absalom-Wong, based on Our World in Data/UN

Two big unknowns cloud the longer-term outlook, says Nick Parr, honorary professor of demography at Macquarie University in Sydney. The first is how quickly fertility rates decline in sub-Saharan Africa—the change has been slow and uneven, and past predictions of decreases have been proven wrong.

The second unknown stems from the end of China's one-child policy, which lasted from 1980 until 2016. 'The United Nations projections assume quite a substantial recovery in the birthrates for China,' says Parr. 'So far, there hasn't really been a recovery, despite the move from a one-child policy to a two-child policy and now to a three-child policy.'

Meanwhile, people are living longer. A baby born in 1900 was expected to make it to 32, on average, globally. The figure was 72.8 years in 2019 and the UN predicts it will rise to 77 by 2050. Longer life spans and fewer babies mean the world's population is ageing. Saikia says that by 2050 there will be twice as many people aged over 65 as children aged under five. Today the median age of a person on Earth is 31.5 years. That will climb to 42.3 years by 2100, the UN forecasts. Age demographics vary considerably around the world's regions. The median age in Europe is now a middle-aged 42.5 years, while in sub-Saharan Africa it is a youthful 18.7 years.

WHICH PLACES WILL HAVE THE MOST PEOPLE BY 2100?

More than half of the world's people live in cities, a number that is expected to jump to two-thirds by 2050. UN projections for individual cities extend only as far as 2030, but they signal longer-term trends. In 1970, New York, Los Angeles, London and Paris all ranked among the world's

10 most populous cities; by 2024, no city in North America or Europe was on that list. By 2030, Tokyo, currently the world's largest city with 36.6 million people, will be overtaken by India's capital, Delhi, which has 29 million inhabitants now but is projected to have 39 million by the end of the decade. Another Indian city, Mumbai, will be in sixth place. Shanghai will be third and Beijing seventh. The newest top 10 entry by 2030 will be Kinshasa in the Congo. (Nigeria's Lagos will rank 11.)

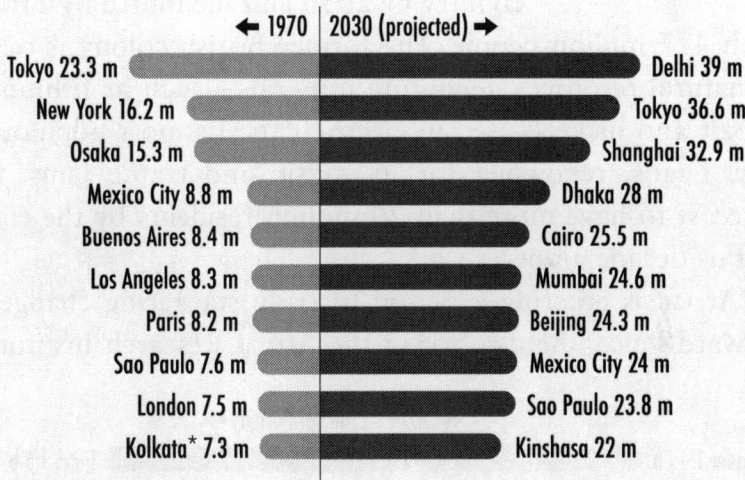

← 1970	2030 (projected) →
Tokyo 23.3 m	Delhi 39 m
New York 16.2 m	Tokyo 36.6 m
Osaka 15.3 m	Shanghai 32.9 m
Mexico City 8.8 m	Dhaka 28 m
Buenos Aires 8.4 m	Cairo 25.5 m
Los Angeles 8.3 m	Mumbai 24.6 m
Paris 8.2 m	Beijing 24.3 m
Sao Paulo 7.6 m	Mexico City 24 m
London 7.5 m	Sao Paulo 23.8 m
Kolkata* 7.3 m	Kinshasa 22 m

*Calcutta became Kolkata in 2001

The most populous cities in the world, 1970 versus 2030 (projected).
Simon Rattray, based on United Nations data

As for nations, in 2023, India overtook China as the most populous nation. It will still have the most people in 2100 (1.5 billion compared with 1.45 now) but China will be a much more distant second (from 1.41 now to 633 million). In the medium term, more than half of the world's projected population increase between now and 2050 will be concentrated in just eight countries: India, the Democratic Republic

More than half of the world's ... population increase between now and 2050 will be concentrated in just eight countries. of the Congo, Egypt, Ethiopia, Nigeria, Pakistan, the Philippines and the United Republic of Tanzania.

Five of those countries are in sub-Saharan Africa, which comprises 46 of Africa's 55 nations. By 2050, the region will be home to a quarter of the world's people, up from 17 per cent today. Its largest economy, Nigeria, is projected to become the world's fifth most populous country by 2050 and the fourth by 2100 with 477 million people. The former British colony is rich in natural resources, including minerals—such as lithium, cobalt and nickel—used in electric cars. Its most populous city, Lagos, renowned for its music and traffic jams, is forecast to have more than 20 million residents by the end of this decade alone.

'Africa is entering a period of truly staggering change,' Edward Paice, the director of the Africa Research Institute

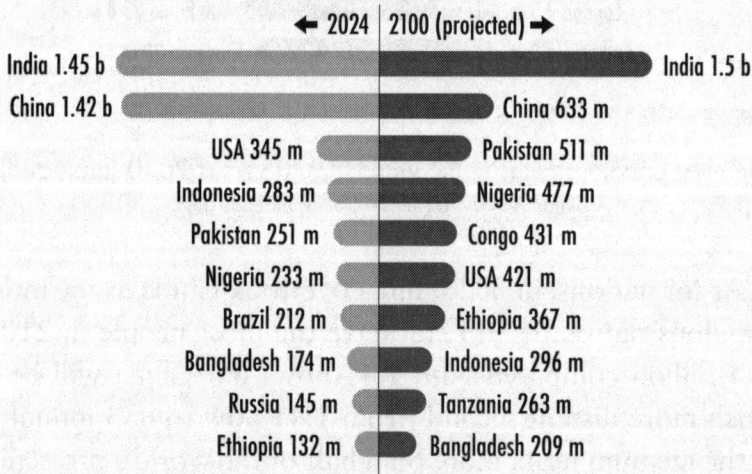

← 2024	2100 (projected) →
India 1.45 b	India 1.5 b
China 1.42 b	China 633 m
USA 345 m	Pakistan 511 m
Indonesia 283 m	Nigeria 477 m
Pakistan 251 m	Congo 431 m
Nigeria 233 m	USA 421 m
Brazil 212 m	Ethiopia 367 m
Bangladesh 174 m	Indonesia 296 m
Russia 145 m	Tanzania 263 m
Ethiopia 132 m	Bangladesh 209 m

The most populous nations in the world, 2024 versus 2100 (projected).
Simon Rattray, based on United Nations data

in London, told *The New York Times*. In fact, by 2100, sub-Saharan Africa is forecast to be the world's most populated region with more than 3 billion people. Meanwhile, Europe, North America, South America and East and South-East Asia are all expected to have fewer people than they do now.

And Australia? Our population stood at 27.3 million in mid-2024. The latest UN forecasts show that will climb to around 33 million by 2050 and to 43 million by 2100.

WHY AREN'T PEOPLE HAVING BABIES?

One reason is the growth in educational opportunities, particularly for girls. One example: in Angola, according to *The Economist,* women with a tertiary education have an average of 2.3 children compared to 7.8 children born to women without any schooling.

Wealth probably also plays a role, as the richest nations have the lowest fertility rates, but its effect is a little more complicated. Being truly wealthy means you can afford more children and have access to a support network (just ask Elon Musk, father of 12). Yet being relatively wealthy but still struggling to keep up with the Joneses can make having children a lower priority in cities such as Melbourne, Sydney, Seoul or Tokyo where real estate can be prohibitively expensive.

Other reasons tend to be more particular to each nation. Take Brazil, where the fertility rate dropped from around six children per woman in 1960 to 1.6 in 2021. A 2012 study found that the viewing of TV soap operas, known as *telenovelas,* was linked to 'significantly lower fertility'. The authors suggested this was largely because the shows' main female characters were typically childless or had only one child.

Hungarian President Viktor Orban has blamed climate science—and fear of Armageddon—for at least some of his country's low birthrate. Despite aggressive government incentives, such as low-interest loans and income tax exemptions for people with four or more children, the country's fertility rate remains stable at a shade over 1.5 children per woman.

South Korea has one of the lowest fertility rates. The IHME predicts it will be one of 23 countries where the population will drop by half between 2017 and 2100. Others include Thailand, Italy and Spain. Another 34 countries are forecast to have population falls of between a quarter and a half, including China (down 48 per cent). So concerned is the South Korean government that it sent a delegation to study the southern Japanese town of Nagi, where childcare is inexpensive, parents get payments for every year their child remains in high school and a job-matching system helps part-time workers balance work and family life.

Social attitudes are harder to shift. In South Korea, experts point to a growing divide between men and women. An 'escape the corset' movement that developed in 2018 and saw some Korean women cut their hair short and ditch makeup, then escalated into the 'four nos': no to dating, heterosexual sexual relations, marriage and childbirth. South Korea also has a hyper-competitive education system and some decidedly family-unfriendly laws—restaurants, museums and even the national library can bar children from entering. A 2023 survey of 15,000 Koreans aged 19 to 34 found barely half of women intended to have children while 70 per cent of men said they would be willing. 'The birth strike is women's revenge on a society that puts impossible burdens on us and doesn't respect us,' Seoul office worker Jiny Kim, 30, told *The New York Times*.

The UN says more family-friendly policies, including more generous parental leave and affordable childcare, will be needed worldwide to build 'demographic resilience'. Italy, whose fertility rate dropped to 1.24 in 2020, recently increased benefits, particularly for families with more than three children, and extended maternity leave.

As the problem becomes even more pressing ... some nations might resort to coercive policies.

The evidence that such polices work is mixed. Poland introduced a monthly benefit in 2016 to encourage families to have two or more children but the country has yet to see an increase in the birthrate, reports *The Economist*. Singapore offers paid maternity leave, childcare subsidies, tax relief and rebates, one-time cash gifts and grants for companies that implement flexible work arrangements yet its fertility rate dropped to 1.12 in 2022.

As the problem becomes even more pressing, the IHME study warned, some nations might resort to coercive policies. 'A very real danger exists that, in the face of declining population, some states might consider adopting policies that restrict female reproductive health rights and access to services,' it concluded. 'Low fertility in these settings might become a major challenge to progress for females' freedom and rights.'

AREN'T FEWER PEOPLE GOOD FOR THE PLANET?

British economist Thomas Robert Malthus made some bold predictions in his 1798 treatise *An Essay on the Principle of Population*, not least that human beings would reproduce exponentially to the point of starvation. The world's population at that time was poised to reach one billion. Indeed, if it had grown as Malthus had warned, we would be overrun today. 'The power of population is so superior to the power of the Earth to produce subsistence for man, that premature death must in some shape or other visit the human race,' Malthus wrote. Yet, as the rather less intellectual but possibly more insightful US baseball player Yogi Berra intoned, 'It's hard to make predictions, especially about the future.'

A decline in global population is potentially good news for the environment and the climate. Fewer people could lower carbon emissions, reduce competition for natural resources and ease pressures on global food supplies.

A dramatic shift in demographics will have ... sweeping, and often negative, consequences.

But while the overall number of people is one matter, a dramatic shift in demographics will have other sweeping, and often negative, consequences. The size of a nation's working-age population is a key economic driver, along with productivity growth and workforce participation. Japan, where a quarter of the population is over 65, offers an extreme snapshot: economic stagnation, emptied-out villages and workers delaying retirement into their seventies. 'We are approaching an era of acute labour shortages in much of the world—and robots are not yet ready to come to our assistance,' demographer Paul Morland said in 2022.

Ageing is creating a 'population bomb', write David Bloom and Leo Zucker for the International Monetary Fund. 'The potential consequences of inaction are dramatic: a dwindling workforce straining to support burgeoning numbers of retirees, a concomitant explosion of age-related morbidity and associated healthcare costs, and a declining quality of life among older people for lack of human, financial and institutional resources.'

The age when people qualify for the old-age pension has been creeping up in many countries as governments attempt to lift workforce participation, which may be fiscally sensible but is rarely popular. When French President Emmanuel Macron pushed through a change to pension eligibility from 62 to 64 (for those born after 1967) in 2023, he triggered widespread protests.

In Australia, workforce participation among people aged 65 and above has risen steadily over the past two decades, making that group an increasingly important contributor to the economy. Compulsory superannuation contributions also mean that a growing share of retirees are self-funded rather than relying on government pensions. Even so, an ageing population poses long-term budget challenges.

WHAT ELSE DO THESE TRENDS MEAN FOR AUSTRALIA?

'When our grandchildren become the policymakers, they are likely to be talking about how to increase population rather than getting worried about a rising population,' says Saikia. He predicts many nations will rely more on immigration to stave off the damaging effects of a population decline. 'Migration sending countries will play a vital future role in the economies of other countries, especially developed countries including Australia,' he says.

Even after Australia's fertility rate fell below the 2.1 replacement level in the 1970s—it hit an historic low of 1.63 in 2022—our population has grown steadily thanks to strong immigration.

Net overseas migration—the number of people who move to a country minus the number who move away—has accounted for more than 60 per cent of Australia's population growth during the past decade, and that number is forecast to reach 75 per cent by the 2060s. The federal government's 2023 Intergenerational Report has projected that the national head count will rise from 26.5 million to 40.5 million by 2063, an increase of 53 per cent.

The IHME study says Australia's liberal migration policies and ability to attract new arrivals will allow the nation to maintain a relatively stable working-age population and 'various economic, social, and geopolitical benefits' that come with it for the rest of the century. 'Nations that sustain their working-age populations over the long term through migration, such as Canada, Australia, and the USA, would fare well.' It forecasts Australia will climb the global economic rankings as a result. In 2017, Australia was the world's 12th-largest economy, says the IHME paper, and it will rise to 11th largest in 2050 and eighth largest in 2100.

Immigration policy has long been a subject of fierce debate in Australia. A temporary immigration boom following the COVID-19 pandemic stoked public anxiety about housing supply shortages and pressures on infrastructure and essential services. In 2023, the Albanese government announced changes to bring 'migration back down to sustainable, normal levels' including a lower intake and more stringent rules for overseas students.

'Nations that sustain their working-age populations through migration ... would fare well.'

Striking the right balance on immigration will challenge Australia's political leaders as the dynamics of the global population shift, even if our numbers are a small part of the planet's story. After all, we account for about 0.33 per cent of the world's people.

17

WHY DO HONEY BEES DO A 'WAGGLE DANCE'?

Bees play a big role in our world despite facing many perils. How does a real 'hive mind' work?

Billie Eder

A ustralia's largest movement of livestock happens in the cool of night. When crops have flowered and are ready for pollination, honey-bee hives are loaded into netted trucks. The precious cargos include queen bees, the linchpins of any hive, and thousands upon thousands of worker bees that serve them.

Humans have turned bees into extremely efficient biological machines. Without them, we wouldn't have almonds, blueberries and a lot of other nuts, fruits and vegetables we enjoy. But these remarkable creatures are also buffeted by the vicissitudes of modern life: destruction of native habitats, exposure to pesticides, extreme weather and attacks from pests such as the varroa mite.

Why does it matter? How are bees faring? And what on Earth is the so-called 'waggle dance'?

HOW DO BEES LIVE?

With its glittery blue and black abdomen, the blue-banded bee flies solo and nests in sandstone banks. Like most of Australia's 2000 or so native bee species, the blue-banded bee doesn't produce honey. Just a handful do: one such is the sugarbag bee, which is small and black and doesn't sting but bites, builds little resin pots inside tree hollows and fills them with honey (as Indigenous Australians have known for many thousands of years).

A hive has up to 40,000 bees, which have some of the most advanced forms of communication among social insects.

But the most common bee in Australia, and the one that makes much of the honey you buy at the supermarket, is a European import. The *Apis mellifera* (Latin for 'honey-bearing bee'), or

Western honey bee, is said to have been transported to New South Wales on the convict ship *Isabella* in 1822. A hive has up to 40,000 bees, which have some of the most advanced forms of communication among social insects, says Dr Cooper Schouten, director of the Honey Bee Research Lab at Southern Cross University in Lismore, NSW. 'Honey bees communicate using bumps, grunts, dancing and pheromones, which are involved in almost every aspect of the honey bee colony life, from reproduction to foraging, defence, orientation and the whole integration of colony activities, from foundation to decline.'

On a typical day, the males (drones) eat honey and prepare to mate with a queen. The workers, who are female, perform various roles to keep the hive in smooth working order: some forage for pollen (bees can visit up to 5000 flowers a day), some use secreted wax to build honeycomb cells, others feed the babies (larvae), while others attend to the queen, and so on. The queen might seem the captain of all this industry, but a beehive is one of the 'purest forms of democracy', says Schouten. 'All of the bees work together to collectively find information and then make decisions based on that information. They will even decide collectively to kill the existing queen and make a new one if she's not performing.'

If the last part sounds a bit like *Games of Thrones*, consider the plight of the queen's potential replacements. If a queen dies, the workers will choose some larvae and feed them highly nutritious royal jelly until one emerges—at which point they will kill the others.

When the new queen takes her maiden flight, she is inseminated by up to 15 drones. To avoid inbreeding, this aerial copulation takes place outside the hive, in an area where drones from many colonies congregate, waiting for a virgin queen to arrive. Together, these drones provide her with enough sperm to last her lifetime; they die in the act of mating. The queen returns to her hive and lays up to 2000 eggs a day. 'She lays eggs all day long—as much as her own body weight each day,' says Schouten. 'And then the worker nurse bees feed her, tend to her, keep her clean and look after her.' While worker bees may live for only eight weeks, a queen can live for up to five years.

Bees are also involved in another form of reproduction: that of plants. Flowering plants produce pollen, which is rich in protein, and nectar, which is a carbohydrate. Both are an important source of nutrition for bees. Bees suck nectar in their proboscis, a kind of hollow tongue, and store it in a sac called a crop. Honey bees collect pollen by stuffing it into pollen baskets called corbiculas, on their legs.

Pollen is produced by male flowers, or male parts of flowers (stamen), and as they go on their collection runs, bees transfer it to a female flower, or female parts of a flower (pistil)—where fertilisation occurs. Only then do many plants start growing their fruit, vegetables or nuts. For honey bees, it's the job of scout bees to discover sources of nectar and pollen, and sometimes even to find new places for the hive to nest. A scout bee will share this information with other bees at the hive through a unique form of communication: the waggle dance.

To the untrained eye, it might look like a bee is shimmying. In fact, it's sharing a code.

WHAT'S A WAGGLE DANCE?

Austrian scientist Karl von Frisch won the Nobel Prize in 1973 for discovering the waggle dance in honey bees. To the untrained eye, it might look like a bee is shimmying. In fact, it's sharing a code.

The dance consists of a series of pathways that look a bit like a figure of eight, the bee makes an arc to one side, then turns and waggles her abdomen as she proceeds down a centreline (called a waggle run) before arcing the other way and waggling back down the centreline, alternating left and right after each waggle run.

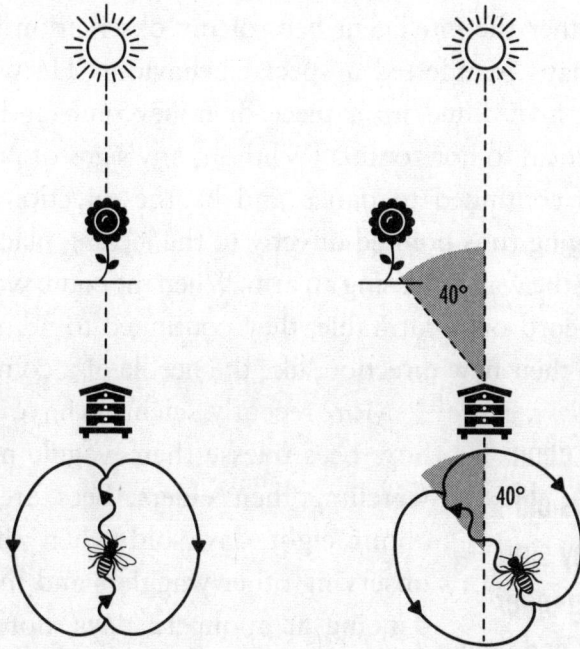

Bees in a hive 'waggle' in the direction of a food source, relative to the sun. *Simon Rattray*

The direction of the centreline conveys where food is in relation to the hive, relative to the sun. If food is in the same direction as the sun, the bee will waggle up the honeycomb in its hive in a straight line; if food is in the opposite direction to the sun, it will waggle straight down. Or it will waggle on an angle to the sun, left or right, if that's where the nosh can be found. The duration of each waggle run conveys how far the bees will need to fly; in general, every second the bee waggles along the line is equal to about a kilometre, says Schouten. The enthusiasm of the dance tells the other bees how good the pollen is. Abundant sweet nectar or high-fat pollen, and the bee will dance its little pollen socks off.

'It is conceivable that some people will not believe such a thing,' von Frisch said during his Nobel lecture. 'Personally, I also harboured doubts in the beginning and desired to find out whether the intelligent bees of my observation hive had not perhaps manifested a special behaviour.' He went to a different hive, lifted up a piece of honeycomb and tilted it from vertical to horizontal. '[W]ithout any signs of perplexity, the bees continued to dance and by the direction of their tail-wagging runs pointed directly to the feeding place, just as we show the way by raising an arm. When the comb was turned like a record on a turntable, they continued to adjust themselves to their new direction, like the needle of a compass.'

Bees are about eight days old when they start observing other wagglers and they begin dancing at about 12 days.

More recently, scientists have observed how bees finesse their waggle moves by watching their elders. Bees are usually about eight days old when they start observing other wagglers and they begin dancing at about 12 days, noted scientists from China and the United States in 2022. The scientists created new colonies of day-old bees but gave only some of them access to older dancers. Bees who lacked guidance from their elders 'produced significantly more disordered dances with larger waggle angle divergence errors and encoded distance incorrectly', the scientists reported in the journal *Science*. With experience, the bees were able to iron out some errors, but their incorrect distance coding was 'set for life'. Bees who followed the example of experienced dancers 'showed neither impairment'.

WHAT MAKES BEES SO USEFUL TO PEOPLE?

When a forager honey bee returns to its hive with the nectar it has collected, it passes it over to what are known as house bees, who chew it, ingest it and regurgitate it from one house bee to the next, with enzymes from the bees breaking down the sugars and proteins into a kind of watery honey. To 'dry' this honey, the bees place it in honeycomb cells where some of the liquid evaporates, and they also fan it with their wings. Honey is a 'super-saturated' solution of complex sugars with some water. 'It is high in antioxidants and can have anti-inflammatory, antibacterial and antimicrobial properties,' says Schouten.

Once the honey is 'ripe', the bee caps the honeycomb cells with wax. Sealed in like this, honey can last indefinitely. Crystallised honey is not spoiled, despite its appearance, Schouten tells us. 'There's simply not enough water in honey to keep all of its sugars dissolved permanently,' he explains. 'Little particles such as pollen, beeswax, bee glue and other nutrients are part of the reason that raw honey is more likely to encourage the formation of crystals.' And if you want to return your honey to its natural state? 'You can simply put your jar of honey in warm water for 30 minutes and stir it,' Schouten advises.

Why are the cells hexagons? The shape allows for the maximum volume of honey to be stored using the least amount of building material. 'If they were any other shape, they'd be wasting wax,' Schouten says.

Australia has about 1800 commercial beekeepers who manage between 400 and 800 hives each on average,

Why hexagons? The shape allows for the maximum volume of honey ... using the least amount of building material.

producing a total 37,000 tonnes of honey each year, says AgriFutures Australia, a government agency. Birds, bats, butterflies, the wind and even some flies can help with pollinating crops. So can both native Australian and feral European bees (which originated in managed hives but now live in the wild). But it is the commercial European honey bees that are trucked around to service citrus fruit, avocados, canola and other crops as they flit from blossom to blossom. 'The European honey bee is a workhorse. We raise it like we do livestock,' says Dr Tim Heard, a former CSIRO research scientist, adding that people 'just kind of take it for granted'.

SO WHAT COULD POSSIBLY GO WRONG?

Varroa mites (*Varroa destructor*) are the world's most devastating honey-bee pest. The red-brown parasite sneaks into brood cells housing larvae and lays eggs. The newly hatched mites then latch on to larval and adult bees, and feed off their blood. 'It would be like you or me having a smartphone-sized tick on us,' says Schouten. The mites also spread viruses to bees, much as mosquitoes spread malaria, which can cause colonies to collapse as worker bees flee infected hives.

Australia had long tried to keep varroa mites out of the country but they are fiendishly hard to police. That became evident in 2022 when the mites were discovered in surveillance hives at the Port of Newcastle, leading to 47,000 hives being destroyed in New South Wales. The following year, it was decided it was no longer feasible to eradicate the mite in Australia—the focus would shift to minimising its impact. For commercial beekeepers, this means monitoring and testing hives regularly and treating any that are infected.

In areas affected by varroa in NSW, some hobby hives have been destroyed. As it spreads, the mite is expected to take a toll on feral European bees, too. 'It's likely that varroa mite will sweep through like a bushfire ... [and] take out those wild European honey bees,' says Shane Hetherington, the Department of Primary Industry's chief plant protection officer.

The absence of the varroa mite until now helped Australia avoid the colony collapses that decimated hives overseas in the early 2000s. In the United States, the crisis put a spotlight on bees, says Dr Judy Wu-Smart, an entomologist at the University of Nebraska-Lincoln. 'Colony collapse disorder was definitely this phenomenon that went through the media and social media, really bringing public awareness to the plight of bees and the importance of bees.'

Wu-Smart, who runs the university's Bee Lab, says bees are a delicate indicator of problems in their environment. She witnessed this phenomenon in her own backyard in Nebraska. A leaking ethanol plant in her area caused 'multiple years of die-off' in her hives. 'My bees were a red flag that there was something seriously wrong in the landscape,' she says.

Insecticides, fungicides and herbicides on flowers are another hazard—bees can collect the poison as they forage. And floods, bushfires and drought are just as dangerous for bees as they are for other creatures. Several thousand hives were lost in floods in northern New South Wales in 2022. Droughts are especially harmful; bees use a total two litres or so of water a day to thermo-regulate a hive. But it's clearing land that is one of the greatest threats to native bees, says Heard. The loss of green corridors—in areas ranging from deserts to rainforests—reduces both food sources and places for nesting.

In 2021, researchers in Argentina, the United States and Germany published a paper in *One Earth* warning of a global decline in bee species. The researchers used the Global Biodiversity Information Facility, a database of life on Earth that draws on museum specimens, scientific papers and even smartphone photos. Since the 1990s, the number of bee sightings has dropped sharply, the researchers found.

That's the broad context. But the insect kingdom has about 20,000 described species of bees, which makes it hard to generalise about their future. Some are endangered, others aren't, and for most species we simply don't know. Australia could have as many as 2500 species of native bees but, according to Taxonomy Australia, many of them could become extinct before they are even discovered. Benjamin Oldroyd, emeritus professor of behavioural genetics at the University of Sydney, says our limited knowledge of Australian native bees is hindering our understanding of what threatens them. 'There have been extinctions but no one really knows how many, and it is tragic because they are keystone species,' he tells us.

WHAT WOULD HAPPEN IF BEES DIED OUT?

'If the bee disappears from the surface of the Earth, man would have not more than four years to live.' This quote is often attributed to Albert Einstein but 'there is zero evidence that Einstein had an interest in bees', says Dr Tobias Smith, founder and director of Bee Aware Brisbane. Regardless of who said it, the quote overstates the case. If bees became extinct, 'the planet would change and we would have issues but

> 'If bees became extinct the world would become a much more inhospitable place.'

probably humanity wouldn't die', Smith says. 'But we would go through enormous change. The world would become a much more inhospitable place but we would still eat all these things that don't require pollination.'

About a third of the food we eat is derived from plants that rely, to a greater or lesser extent, on pollination, says Oldroyd. If bees disappeared, we could still produce crops such as wheat, corn and rice but there would be a 'reduction in the food supply and the food quality, and some crops would be decimated', he adds. Almonds, with their blossoms, are 'the poster child because they 100 per cent rely on bees'.

At the very least, if bees were to die out, 'we will feel a crunch in our wallet', says Wu-Smart. 'That can lead to human health issues because healthier foods are less accessible.'

With the world's population continuing to grow, at least for the time being, we need bees more than ever, notes Tim Heard. 'We don't really understand what the full outcome of [bee extinction would] be,' he says. 'It's too important for us to take risks with.'

The United States and other countries have looked into using robotic bees for pollination but, for now, the technology is costly. While research is underway into genetically modified 'Frankenbees', they are yet to be viable and even if they were, many researchers worry that they could disrupt the natural balance of ecosystems. Meanwhile, small drones have been trialled in greenhouses in Australia where they vibrate the flowers of strawberry and tomato plants to trigger pollination.

We can help bees in our own gardens, says Smith. Weed by hand rather than using chemical weed killers. Plant edible herbs and let them flower. If you have space, plant a range of native plants 'because different bees have different tastes'.

And find out what type of bees are in your area (the blue-banded bee, for instance, is partial to blue and purple flowers). 'Bees are very cute, fascinating little insects and they also hold everything together,' Smith says. 'Once you know them, you can't give up on them.'

18

WHY ARE TINY MICROCHIPS AT THE CENTRE OF A GLOBAL TUSSLE?

The chips in our phones, cars and weapons systems are devilishly tricky to make. Who will get the edge in this vital and secretive multi-billion-dollar industry?

Eryk Bagshaw, Jackson Graham and Daniel Chen

The factories run 24/7. Outside, trees obscure the gas tanks that keep them running. Guards patrol the perimeters and ask us to delete photos of the buildings. Inside all of them, finely calibrated machines perform exquisitely finicky tasks on a nanoscale. Canteens work overtime to feed thousands of specialist employees.

The tiny pieces of technology these factories are making—advanced semiconductors, or microchips—are at the centre of a struggle between the United States and China that will help define the 21st century. The chips, made of parts tinier than viruses, are essential to the phone in your pocket, your car and your microwave; they power not only the latest generation of smart homes but also artificial intelligence and military equipment like fighter jets.

The tiny pieces of technology are at the centre of a struggle between the US and China that will help define the 21st century.

Some 90 per cent of the world's most advanced chips, including those used in the West, come from factories a few hundred metres apart in Hsinchu, just over an hour's drive south-west of Taipei. The area is known as Taiwan's Silicon Valley. Fierce competition among the tech giants, fuelled by an insatiable demand for chips, keeps these sites running hot. 'It's winner takes all,' says Robert Tsao, the billionaire founder of one of the companies, United Microelectronics Corporation (UMC). But the industry, he says, needs to be highly protected from geopolitical tensions. 'Otherwise, it could become a hostage.'

Making these chips is eye-wateringly expensive and devilishly difficult, and there are pinch points along their supply chain. So essential is Hsinchu to the global technology industry that even a short shutdown—caused by, say, an earthquake—can put the industry on edge. Taiwan is earthquake prone. Production was paused and some chip plants were evacuated after a 7.2-magnitude quake in 2024, which led analysts to point out that even a 'hiccup' could delay shipments and cause millions of dollars in unexpected costs.

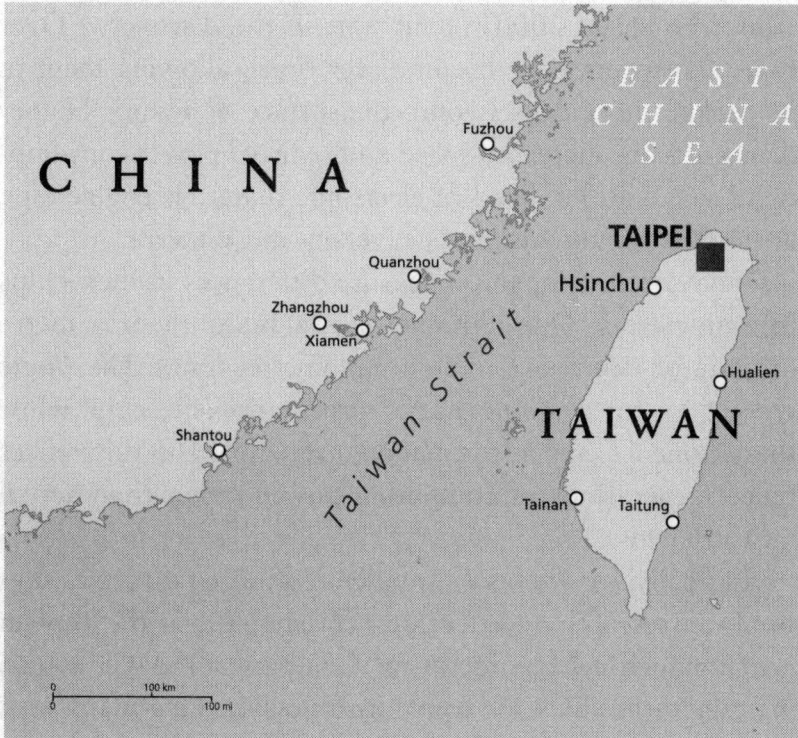

Most of the world's most advanced chips come from Hsinchu, Taiwan's 'Silicon Valley'. *Peter Hermes Furian/Shutterstock*

What happens if these chips fall into the wrong hands? Why is there a race to produce them? Why have they become a flashpoint?

WHAT'S A CHIP?

In 1956, three US scientists won a Nobel Prize for inventing transistors, tiny switches (then a centimetre long) that improved on valve technology. They quickly became a mainstay of electronics. Two years later, engineers Jack Kilby and Robert Noyce realised that electronic systems could be enhanced when many transistors were arranged on what's called an integrated circuit, or microchip. (Kilby won a Nobel in 2000 for his part in the discovery.) Over time, transistors have become ever tinier, allowing them to be packed more densely onto the surface of a chip. 'Higher density means increased speed and reduced power consumption,' says Lan Fu, head of electronic materials engineering at the Australian National University in Canberra.

Today, tech companies measure differences in their chips in nanometres. One nanometre is a billionth of a metre (a grain of rice is five million nanometres long). The Apple iPhone 15, for example, has a 'three-nanometre chip' while the iPhone 14 has a 'five-nanometre chip'. The differences reflect generations of computing power rather than actual measurements.

'Basically, we are faster, smaller and more powerful,' says Bobby, a research and development engineer at the Taiwan Semiconductor Manufacturing Company (TSMC), which manufactures chips for Apple in Hsinchu. (Like many tech workers who spoke to us, Bobby asked to be identified with a pseudonym to protect his employment.) China, which lags behind Taiwan, has been pumping billions of dollars into its largest chip producer, the Semiconductor Manufacturing International Corporation (SMIC). Its goal is to produce five-nanometre chips.

Microchips can have up to 100 layers. A Dutch company that supplies chip manufacturers, Advanced Semiconductor Materials Lithography (ASML), likens them to tiny sky-scrapers, thanks to their three-dimensional grids of billions of transistors, which switch on or off to process the ones (on) and zeroes (off) of computer code. Manufacturing microchips is an elaborate process. High-quality quartzite is purified by heating it to 1500 degrees Celsius and then put through a second heating process to make super-pure poly-silicon, says Ed Conway, author of *Material World*, a book about six of the substances that will shape the future. 'The whole point of silicon chips is they need to be incredibly pure,' Conway tells us from London. 'A single rogue atom in the wrong place means that your electrical current can't go through and, therefore, your silicon chip fails. It is one of the purest things that humankind is capable of making.'

'A single rogue atom in the wrong place means ... your silicon chip fails. It is one of the purest things that humankind is capable of making.'

The second stage of heating is done in a crucible, a bowl-shaped container made from a special type of purified quartzite that only one mining district produces in large quantities: Spruce Pine, North Carolina. 'Pretty much every microchip in the world will have come into contact, shall we say, with a Spruce Pine crucible,' Conway says. 'If this place goes down, we're all in big trouble.'

The silicon cools to form a sausage-like shape, which a diamond saw slices into circular wafers that go to fabrication plants, or fabs. Taiwan has the lion's share of fabs that produce 'logic' chips—the technology that runs smartphones and computers—while South Korea's facilities build more than half of the world's memory chips, which store data. The fabs have machines that imprint the shapes of the tiny transistors on each wafer, a process called photolithography. Put simply, a machine shoots light at sensitive chemicals to make patterns on the silicon.

Most of the key chemicals in this process come from Japan, and the most advanced of the imprinting machines are made by ASML. 'If there were a vast flood in the Netherlands that knocked out that production facility, the machines that make chips would be delayed, and therefore chips would be delayed,' says Chris Miller, author of *Chip War: The Fight for the World's Most Critical Technology*.

WHY IS THE CHIP BUSINESS SO SECRETIVE?

The fabs themselves are another world, says Miller, an associate professor of international history at Tufts University in the United States. People wear hazmat suits in pristine surrounds—a speck of dirt can ruin a silicon chip. 'It's one of the least human environments you can be in,' he tells us from New York. 'It's just mostly robots, and the people who are there are completely covered up so you can barely make out who they are. All of the manufacturing is happening at a scale that is far smaller than the human eye [can see].'

Conway says part of the success of the leading fabs is that they can ensure a high percentage of their transistors will work properly. 'What that underlines is being good at making chips . . . is not just about having the right tools, the right machines and the right buildings, it's about running manufacturing plants,' he explains. UMC founder Robert Tsao put it like this when we spoke with him in his office in Taipei: 'The amount of money needed to upgrade technology and equipment is huge. For a high-end semiconductor used in a car, making one piece will take half-a-billion dollars. So unless you're already in mass production, you just can't do it yourself—it will be too risky. Maybe even crazy.'

All of this investment has made the industry highly secretive. No personal devices are allowed inside TSMC, for example. Each employee is issued with a company phone that's had its photo and storage capabilities disabled. It has only one ringtone. On weekends, when a company phone rings at a public place like Hsinchu's Big City shopping centre, many employees 'on call' will check their pockets to see if it's theirs. 'It is high pressure,' says Steven, a mask data integration engineer at TSMC. 'We make products for major clients such as Apple. I joke with my friends that maybe [they] can buy more iPhones so that my salary will increase.' He says the maximum number of errors allowed in chip production is three per day; any more and workers' pay can be docked. Performance-based bonuses make up two-thirds of wages.

A TSMC spokesperson said the company's annual employee turnover rate was a relatively low 6.7 per cent and that it encouraged staff to maintain a work–life balance. 'TSMC is in one of the most competitive industries in the world,' the spokesperson said. 'In recent years, we have pushed our technology to a world-leading position. It is not easy to maintain this position, but the company believes that while hard work is necessary, work is only one part of life.'

Bobby, who works in research and development at TSMC, says the pressure is worth it. Taiwan's universities churn out specialised graduates who are hoovered up by the chip companies, future-proofing the industry. Employees fresh out of uni earn $70,000 a year while employees with a PhD take home $100,000—far more than Taiwan's average annual wage of $33,000. 'The technology industry is an oasis in a generally low-paid country,' Bobby tells us.

WHY IS THERE SO MUCH GEOPOLITICAL TENSION OVER CHIPS?

Chips are often described as 'the new oil' and the comparison isn't far off the mark, says Greg Allen, an expert in advanced technologies at the Centre for Strategic and International Studies in Washington, DC. Much of the production of the crude black goop is concentrated in a few countries, including the United States, Saudi Arabia and Russia, yet demand for it is global. 'In the case of semiconductors, this used to be an industry that didn't think a tonne about politics,' says Allen. 'That era is profoundly over; semiconductors are a deeply geopolitically relevant industry.'

A fact sheet issued by the White House in 2024 stated, 'Semiconductors were invented in America and serve as the backbone of the modern [global] economy. But today, the United States produces less than 10 per cent of global supply and none of the most advanced chips.' The United States wants to raise its share of chip production to 20 per cent by the end of this decade. Analysts question whether that can be achieved. 'Part of the reason why Korea and Taiwan occupy the position that they do in the semiconductor industry is they can invest rain or shine,' says Allen. 'It's because these US$20-billion [$30 billion] facilities take two or three years to bring online. And that's the one question I have for the United States in this area: If your goal is 20 per cent, are you really prepared to invest that?'

TSMC is building a $60-billion, four-square-kilometre fab complex in Arizona, which it says will start production in 2025 despite being plagued by delays. The company has also poured $30 billion into Japan, opening its first fab there in 2024 with plans for a second by the end of 2027, and it is looking to build a fab in Germany.

Meanwhile, China's economy has boomed off the back of tech companies such as Huawei, Tencent and Xiaomi. All of them rely, to an extent, on chips made in Taiwan and the West. In 2018, Chinese President Xi Jinping told chip firm Yangtze Memory Technologies Co. that developing chips was as important for China's technology sector as hearts are for humans. 'When your heart isn't strong, no matter how big you are, you're not really strong,' he said.

China's economy has boomed off the back of tech companies such as Huawei, Tencent and Xiaomi. All . . . rely on chips made in Taiwan and the West.

The race to develop artificial intelligence (AI) is not for the faint-hearted. The US military plans to field thousands of 'autonomous systems' run by AI, such as drones, to counter China's rising military might by 2026. 'But it's not only that,' says Chris Miller, the author of *Chip War*. 'There has been, for some time, the application of AI to electronic warfare systems to jam your opponent's communications and keep yours open. As AI gets more sophisticated, so too will electronic warfare.'

China wants to be the leader in AI by 2030. 'Over the next decade, or two or three decades, we expect pretty much everything in military affairs is going to involve artificial intelligence to some extent,' says Allen. As director of the US Defense Department's Joint Artificial Intelligence Center in 2021 and 2022, he was tasked with initiating talks with China about the responsible use of AI. 'We made this request [for talks] multiple times to the Chinese military,' he says. 'And the answer was "no" every time.'

Amid all of this jockeying, something had to give. In 2022, the US government took a major step: it banned the export to China of advanced chip technology used in AI and for military purposes. 'It actively sought to degrade the techno-logical state of the art in China,' says Allen. The subtext of the ban, he adds: 'When you say to China you're no longer going to be able to purchase the most advanced US-made AI chips, you're essentially saying to China, "The United States and our allies are going to the future, the future is going to be powered by AI—and we don't want you to come".'

The bans didn't completely stop China from getting its hands on US chips. In 2023, Reuters reported that buyers could purchase a small number of chips designed by US company Nvidia in electronics districts in Shenzhen, although they cost US$20,000—double the usual price. The journal *Foreign Policy* reported that potentially 40,000 to 50,000 banned chips had found their way into China through neighbouring countries. The United States has since closed some of the loopholes that made these backdoor purchases possible. China's Foreign Ministry spokesman Wang Wenbin said, 'The US needs to stop politicising and weaponising trade and tech issues and stop destabilising global industrial and supply chains.'

The most advanced chips the United States has banned from entering China, Nvidia's graphics processing units (GPUs), were originally developed for video games. 'That is why you've had this intermingling of civilian and military concerns with the chip industry,' says Miller. As demand for chips needed for AI technologies grew, Nvidia became one of the hottest stocks on Wall Street and earned a spot among the United States' most valuable tech companies such as Apple and Microsoft.

Although chips are small and lightweight, the vast quan-tities needed to run AI systems make circumventing the bans difficult, say experts we spoke with. 'Chinese companies are definitely complaining about access to chips,' says Allen.

Miller thinks the bans will slow, rather than stop, China's progress on AI. UMC's Tsao doesn't believe China will catch up. 'Because so much technology has been accumulated outside China,' he says. 'As a newcomer, it's too late.'

Either way, the speed at which Taiwan and the West can outpace China is slowing because of the physical limits of microchips. Manufacturers can fit only so many transistors—even if they are smaller than 1/500,000 of a grain of sand—on a chip. TSMC and other industry leaders are now experimenting with ways to stack more transistors on top of one another. 'It's harder to run faster than the fastest one of the group,' Dan Hutcheson, the vice chair of Canadian semiconductor research firm TechInsights, told *Nikkei Asia* in 2023. 'But once the fastest one can't run any faster, then others catch up to it. In a chip race, the leader cannot afford to trip, even once.'

WHY ARE CHIPS CALLED THE 'SILICON SHIELD' OF TAIWAN?

Up to half of the world's shipping passes through the Taiwan Strait, the narrow strip of water that separates democratic Taiwan from mainland China. Communist China has never ruled Taiwan but Beijing claims the island as its own. According to a 2023 poll conducted by National Chengchi University's Election Study Center, only about 6 per cent of Taiwanese voters were in favour of ever unifying with China, down from 12 per cent in 2018. Undeterred, Xi has repeatedly vowed to unify the mainland and the island 'by force if necessary'.

'They have to think hard about it because it will cause a $3-trillion loss to the world economy,' says Tsao. The 76-year-old has small red and green tattoos on his middle fingers. They read 知 定 or *zhi-ding*, the belief that restraint will pave the way for concentration and, eventually, tranquillity. Tsao's philosophy is Taoist, but his economics is more inspired by Adam Smith, the Scottish economist who wrote *The Wealth of Nations* and laid the foundation for modern free markets based on specialisation.

Taiwan's exports were once concentrated in manufacturing, steel and petrochemicals, but a 1974 breakfast meeting at a local soy milk restaurant changed all that. US-educated electrical engineer Pan Wen-yuan suggested to the minister of economic affairs, Sun Yun-suan, that Taiwan should invest in chip technology, which he had been studying. At the time the Radio Corporation of America (RCA) was making chips for digital watches but its production lines were slow and inefficient. In 1976, the Taiwanese government signed a technology transfer deal with RCA and invested $15 million in the company.

Tsao, an electrical engineering and management graduate from National Taiwan University, was one of the first trainees to travel to the United States to study RCA's chip manufacturing process. The chips were soon being used in much more than watches. Tsao saw their potential value, and the government invested heavily in his company, UMC, as well as in TSMC. The plan worked. By 2020, Taiwan's chip industry accounted for half of all global foundry revenue. Now worth $221 billion a year, it is sometimes referred to as a 'silicon shield' because any threat to Taiwan's chip factories could seriously harm the world economy. China would not be immune from the fallout.

Taiwan's chip industry is referred to as a 'silicon shield' because any threat to Taiwan's chip factories could seriously harm the world economy.

The point was illustrated during the COVID-19 pandemic. A surge in demand for technology from people working at home coupled with restrictions on international shipping slowed the delivery of chips to a crawl. The situation threatened the production of everything from cars and game consoles to medical devices and hospital systems. President Joe Biden's top economic adviser said the chip shortage probably wiped a full percentage point off the United States' gross domestic product in 2021.

The same year, Taiwan's then President Tsai Ing-wen said the country's place in the world's electronic supply chain meant it had to be protected. 'Our semiconductor industry is especially significant: a "silicon shield" that allows Taiwan to protect itself and others from aggressive attempts by authoritarian regimes to disrupt global supply chains,' she wrote in *Foreign Affairs*.

Tsao worries about the country's vulnerability—in 2022 he donated millions of dollars to an academy for citizen defence training. He also fears that the country's chip industry could become less of a shield and more of a hostage in Beijing's negotiations with other countries over Taiwan. 'They might say, "If you keep supporting Taiwan, I will destroy Taiwan's semiconductor industry",' Tsao says.

But some people, including workers in Taiwan's technology sector, aren't convinced of the danger. 'This is the safest place in Taiwan in the event of a Chinese military takeover,' says Jimmy Wang, a 40-year-old semiconductor technology researcher in Hsinchu. 'Why? Because of the concentration of high technology. They want this, all of it. They love this kind of stuff.'

Meanwhile, as new phones, new data centres and ever-more-powerful computers and systems continue to come online, the stakes will only get higher, the competition more intense. 'The chip industry is highly globally interconnected,' says Tsao. 'No country can do it independently of other suppliers. It's a global orchestra and the United States is the conductor.'

19

WHAT'S A NARCISSIST?

Being vain, self-centred or a jerk doesn't make you a narcissist. So what does?

Samantha Selinger-Morris

'That's such a narcissist response,' I overheard a fortysomething man say in a cafe one day. 'I know! It's all about her,' his dining companion replied. 'We had to sell the house!'

At any given moment, it seems, someone is calling someone else a narcissist. 'It's penetrating popular culture to the point where any person you may have had a bad relationship with is a "narcissist",' says Dr James Collett, a lecturer in psychology at RMIT University in Melbourne.

Some people, whether they are in the spotlight or not, may appear to be modern-day versions of Narcissus, the beautiful young man in Greek mythology who loved no one until he saw his own reflection. (He was still single and gazing at himself when he died.) But narcissism is more nuanced than that. 'When someone's clinically diagnosed with narcissism, it's often a very different thing to them just kind of being a jerk,' says Collett.

So, what makes someone a narcissist? What's it like to live with one? And is there a cure?

WHAT MAKES SOMEONE A NARCISSIST?

Collett recalls a counselling session with a man who had lost his job and was having relationship problems. Collett asked him about his sex life—a standard question that can shed light on the state of an individual's relationships. 'He wouldn't shut up about it,' Collett recalls. 'He just kept talking about all these attractive women who want him. "And here's this girl I'm seeing now"—and the guy literally pulls out his phone and starts showing me photos. "Look at that, James. Don't you wish you could get a woman like that?" You just kind of want to cover your eyes—please, please stop!'

Sound like your co-worker or that annoying ex? Not so fast.

Being conceited or boastful is one thing, being diagnosed with narcissistic personality disorder (NPD), as Collett's patient was, is another. Someone with NPD wants 'to establish a dynamic where they're better than you', says Collett. It's a narcissist's 'go-to strategy for every single interaction. It's inflexible across context'. Since 1980, the *Diagnostic and Statistical Manual of Mental Disorders* (known as the *DSM*), has said people who suffer from NPD display a collection of 'maladaptive' personality traits: a pervasive pattern of grandiosity, a need for excessive admiration, a lack of empathy and a sense of 'interpersonal' entitlement. They are exploitative, arrogant and prone to envy. Only about 1 per cent of the global population meet these criteria, says Associate Professor Danny Sullivan.

> Someone with narcissistic personality disorder wants 'to establish a dynamic where they're better than you'.

Sullivan is a consultant forensic psychiatrist and a former director of the Thomas Embling Hospital, Melbourne's high-security mental health centre. 'We all have forms of self-deception but with narcissistic personality disorder, they're much more accentuated and they pervade the person's existence,' he explains. 'They're manifest in their employment, in the way they interact with their partners— they'll always be the person who can do something better than someone else.'

Narcissists are vulnerable, too. Dr Neil Jeyasingam, a psychiatrist at the University of Sydney who has treated people with NPD, describes them as 'basically, like blown-up balloons; you prick them and they will burst'. He continues:

If there's any potential threat to the idea that they're not God's gift to the world, they decompensate horribly.

'A narcissistic personality disorder [sufferer] is one who cannot survive without endless admiration from others. If there's any potential threat to the idea that they're not God's gift to the world, they decompensate horribly. That's the difference pathologically: it's not just about, "Oh look at me, look at me". It's more like, "I look at you and think, *If you are here to worship me, that's good; if you're not, you have no value*".' Jeyasingam prefers the words 'vain' or 'self-absorbed' to describe people who are merely full of themselves.

A series of case studies in *Focus: The Journal of Lifelong Learning in Psychiatry* describe a medical resident who drank 'two to three' bottles of wine to fend off anxiety about an evaluation of his surgery skills—and thought himself 'a real genius' to be able to perform the surgeries while drunk. 'Thinking of himself as an exceptional human being, he believed that his drinking was excusable, if not commendable, that common rules did not apply to him,' reads the report into his treatment for narcissism. (The man sought treatment only at his wife's urging when she noticed he had drinking-related tremors.)

To be clinically diagnosed as a narcissist, the person must be suffering as a result of their personality. 'Everyone has a personality,' US psychiatrist Allen Frances said in 2017 after some mental health professionals in the United States signed a petition claiming then-President Donald Trump was so mentally unwell that he was unfit for office. 'It's not wrong to have a personality; it's not mentally ill to have a personality. It's only a disorder when it causes extreme distress, suffering and impairment [to the person].' Frances, who helped write the *DSM*, argued that although Trump demonstrates 'in pure form every single symptom' of narcissistic personality disorder, he didn't meet the threshold for a diagnosis. 'Trump certainly causes severe distress and impairment in others, but his narcissism doesn't seem to affect him that way,' he wrote. (Frances also said attributing Trump's behaviour to an illness was a 'stigmatising insult to the mentally unwell'.)

Still, many people who don't meet the threshold for an NPD diagnosis can have 'clinically significant' narcissistic traits. Sullivan likens it to being tuned in to a particular radio station, but with the volume turned down. 'They've got personality features which other people recognise [as narcissistic] but the degree of damage or harm is not so great,' he says. 'They're less intractable. They're self-obsessed . . . [but] they probably have a little more insight into the fact that things aren't quite right for them. They've got problematic relationships but can sustain a relationship.'

CAN YOU SPOT A NARCISSIST?

'Oh gosh, no,' says Jeyasingam. 'You'd probably think this was a person that was really nice and really friendly and really keen for me to get to know them.'

In fact, someone with NPD might want to make you one of their boosters. 'People with personality disorders . . . have particular things which are exaggerated, which makes them very interesting and entertaining to watch behave,' says Jeyasingam, who notes narcissistic characters in pop culture such as meth cook Walter White in *Breaking Bad* and the womanising Barney Stinson in *How I Met Your Mother*. 'But [in real life] they can't grow, they can't develop, they can't achieve what their actual

'They can't grow, they can't develop, they can't achieve what their actual potential could be in life.'

potential could be in life. Their capacity for social relationships is limited.'

W. Keith Campbell, a behavioural psychology professor at the University of Georgia in the United States and a leading expert on narcissism, puts prospective study participants into hypothetical social scenarios to determine which ones are high in narcissistic qualities. 'We'll run them through the experiment, [and tell them], "Oh, bummer, no one picked you . . . you failed the test. Is it your fault or do you want to blame the professor? Oh, blame the professor." You set up social interactions to see how people respond,' he says.

In 2013, Campbell and his colleagues studied the 42 US presidents up to George W. Bush to test a hypothesis that 'grandiose' narcissism was a 'double-edged sword' in leaders. They found that presidents who had signs of narcissism were adept at public persuasiveness, crisis management and agenda setting—as well as 'unethical behaviour'. Campbell says Lyndon B. Johnson rated 'high in narcissism'. (Johnson famously pulled down his pants and brandished his wares to a group of journalists after one asked why the United States was involved in the Vietnam War. 'This is why,' he said, as he revealed himself.) 'He made horrible decisions because of ego,' Campbell tells us. John F. Kennedy also rated high. 'You go, "That's a great dude but not a guy you want to marry, necessarily." That's the complexity, though. Everyone loves that priest, loves that physician, loves those politicians—except his freaking wife and kids, or her husband and daughters.'

But not all narcissists are grandiose. Some are distant and aloof and others appear to be modest and unassuming.

It's not uncommon that when they stop receiving external validation, they present to psychologists as profoundly withdrawn and depressed.

WHAT DAMAGE DO NARCISSISTS DO?

'It's actually quite a disability,' says Danny Sullivan. 'You end up with a person whose life can be an absolute train wreck, and it's all their own making. They constantly make poor choices, and they can't relinquish or accept that someone might do something better.'

While no causal relationship has been established between being a narcissist and committing crimes or behaving violently, a narcissist's lack of empathy can lead to conflict—which a narcissist will frame as someone else's problem. 'They will talk about how a person simply failed to acknowledge that they were far more brilliant than anyone in the room, that was why they were abusive towards them,' says Sullivan, who has assessed people for criminal court cases involving fraud and sexual assault. 'People with narcissistic personality disorder are much more prepared to override the rights of others because they so prioritise their own rights— or feel that they are thoroughly deserving.'

'They can come out with some really quite staggering, self-inflated concepts.'

A person with NPD who has, for instance, committed sexual assault might minimise the other person's experience, says Sullivan. 'So they're less likely to be able to appreciate the degree of harm a person's reported. Or they might disregard it as the person lying to advantage themselves— because that's the way a narcissistic person navigates the world, they lie to advantage themselves.' He has seen many narcissists demand 'exceptional treatment' even while in prison. 'Look, they can come out with some really quite staggering, self-inflated concepts: "The ordinary prison psychologists aren't smart enough to deal with my problems, so I'd like you to organise my own psychologist to come in from outside." They'll say, "Well, I'm not like the other ones here. I need a special bed. I need day leave. I need this, I need that." Not because of any specific health-based reasons but simply because they deserve it.'

Often, the culmination is a 'narcissistic crash', after, for instance, a narcissist has lost their job or relationship or their children have abandoned them. In other words, the world stops validating the way they see themselves. 'They're not going to a psychologist and saying, "I think I'm really good, can you help with that?",' says Collett. 'When you see them, it's normally because life has shown them, in a very difficult way, that maybe they're not perfect.'

In the most extreme cases, they may end their own life. Another case study in *Focus* told the story of a 24-year-old research assistant who made two suicide attempts before seeking treatment for narcissism: 'She felt such fear of losing her competence—and hence her reputation in the lab and appreciation of her supervisor—that she saw ending her life as the only way out.'

WHAT'S IT LIKE TO LIVE WITH A NARCISSIST?

Clinical psychologist Tamara Cavenett has seen the partners of narcissists in floods of tears in her office. Narcissists struggle to see another person's point of view, says Cavenett, a former head of the Australian Psychological Society. It's not an 'easily reached-for thinking process'. Where an average person will feel compassion for someone who trips and falls down, she says, a narcissist will 'skip over' that person's point of view because they're more focused on what the impact of someone tripping is on *them*.

To live with a narcissist can be 'quite soul-destroying', Cavenett says. 'Because it's so frustrating, you can't ever get your own needs acknowledged, your own perspective understood.'

Campbell, who wrote *The Handbook of Narcissism and Narcissistic Personality Disorder*, compares being in a relationship with someone with NPD to trying cocaine. 'The trade-off is the same: "I love this cocaine, it's great!",' he says. 'Then, two weeks later, you're like, "I haven't slept in two weeks, I keep listening to disco, I don't know what's going on with me, I'm losing my mind." I'm like, "You shouldn't have done the cocaine!" Cocaine is kind of a liar. It's the same with narcissism. [The narcissist wants] to feel important and special and they're looking at you as a tool for them to become powerful or think they're special or important.

'So if they start dating you, it's the same as them buying a new Porsche. It's fungible. You are like an object . . . You get trashed in relationships if what you want from somebody is affection or love. It's terrifying.'

The children of narcissists also suffer. For instance, if a child is an accomplished athlete, musician or actor, 'these kids get psychologically wrecked', says Campbell. 'Their love is only within this narrow band of performance. If they fail, they lose love.'

Abuse isn't uncommon in relationships with narcissists. 'We can imagine if you are extremely arrogant and self-centred,

> To live with a narcissist can be 'quite soul-destroying, because . . . you can't ever get your own needs acknowledged.'

that's going to be a fairly fertile environment for abuse,' says Collett. Narcissistic abuse can manifest itself as vengefulness, being unforgiving, lying, manipulation, ostracising victims from others or withdrawing communication as a way of resolving conflict.

A narcissist will often impose double standards too, says Sullivan. 'A person will demand loyalty from their partner but consider it appropriate to have extramarital affairs.'

ARE NARCISSISTS BORN OR BRED?

Leading child psychology researchers agree that, as with most mental health conditions, it's both, says Cavenett. 'They think you need both genetics, which creates a limitation on your expression of anything, and then an environment that alters where you fall in a given narrowed spectrum.'

Environments that don't encourage a sense of perspective, don't teach the importance of another person's point of view or that reward superficial, attention-seeking behaviours are more likely to produce a narcissist. 'It's really looking at what behaviours are often being reinforced in the environment,' Cavenett says. 'Think how many times they're told, "You're really successful because you won a medal" versus "How did you behave towards everyone else when you won that medal?".'

Behind the disorder, say experts, is the absence of—or the presence of an extremely damaged—self-esteem. Narcissism is an unconscious defence mechanism that bolsters someone whose sense of self is fragile, according to the prevailing psychological theory. One influential study by a US psychiatrist suggested narcissists experience emotional neglect in their childhood, particularly from their mothers, says

Jeyasingam. 'Harry Stacks Sullivan studied more than 100 young males who had narcissistic personality disorder, and he found almost all of them had maternal issues. The theory is they are essentially ignored and abandoned by their mothers and cope with it by creating an inflated sense of artificial self-esteem.'

But while much of the literature points to men suffering from NPD in greater numbers than women, we shouldn't assume that it's a mostly male disorder. Narcissism might not be as easily detected in women. 'They're more subtle in their interactions,' Jeyasingam says. 'With men, it's "Look at me, look at this amazing tower I built".'

While the literature points to men suffering in greater numbers than women, we shouldn't assume it's a mostly male disorder.

CAN NARCISSISM BE TREATED?

'It's a big deal to get rid of, to treat it,' says Jeyasingam. 'It's not like you throw on a medication and they're better.' Psycho-dynamic therapy, which focuses on the roots of emotional suffering, is the main treatment. In a 1914 paper, 'On Narcissism: An Introduction', Sigmund Freud posits that 'primary narcissism' is a stage of development when an infant is aware of others only as an extension of themselves. Some people, Freud argues, get stuck in this stage and remain unable to love another person as separate from themselves.

If psychodynamic treatment is done properly, the patient gets worse before they get better 'because they feel that core of emptiness, which they have built an entire shell to protect themselves from', says Jeyasingam. 'That's part of the treatment. It's about being able to recognise that core, so you can start to develop again.'

Cavenett has seen people greatly improve with therapy. 'I've certainly had people who've had their relationships back on track, whose relationships—with partners and children—are more engaged.'

Collett is trialling a video program for people with narcissistic traits. The videos show sufferers how to use journals and meditation to disrupt self-critical thoughts and to practise self-compassion. 'So the idea is that this breaks down the need for that wall' and extends participants' focus beyond themselves to other people. It helps them understand that it's all right to have things that don't go well, he says. 'There doesn't have to be a reason or a sense to that, and it doesn't have to mean that you're a failure as a person.'

If you or anyone you know needs support, call Lifeline 131 114, or Beyond Blue 1300 224 636.

20

WHAT TREASURES LIE DEEP BENEATH THE SEA?

Kilometres under the waves, the dark, cold
ocean floor teems with strange creatures.
But as scientists find wonders, miners
are eyeing other riches.

Sherryn Groch

Nine years before the Moon landing, an engineer and a navy lieutenant piled into a submersible chamber the size of a refrigerator to boldly go where no one had gone before—not up to the stars, but down to the deepest point in the ocean. Eleven kilometres down into the Mariana Trench. That's deeper than Mount Everest is tall.

There are mountains under the sea too, and volcanoes that burp out giant lava bubbles. There are armoured spider crabs with legs more than three metres long, 200-year-old sharks and yes, giant squid. Really, the deeper you go, the stranger it gets.

Back in 1960, explorers Jacques Piccard and Don Walsh took nearly five hours to reach the bottom of the ocean. Out the windows of their tiny submersible the pair saw creatures gliding in darkness, even in the deepest waters.

Nearly three-quarters of our planet's surface is covered by ocean yet we have explored only about 10 per cent of it. That's because so much of the sea is *deep*, a dark realm of such extreme conditions that NASA goes there to test the technology it will take to other planets. (The 1960 submersible sent into the Mariana Trench used a clever 'rebreather' life support system later adopted by spacecraft.)

In 2023, the world was reminded of the dangers of exploring this underwater wilderness when tragedy struck: a private submersible journeying to the wreck of the *Titanic*, some four kilometres down in the Atlantic, imploded, killing all five people on board.

Still, today many people are eager to open the deep for business. This sunken scape is home to metals now sought after for electric car batteries, wind turbines and mobile phones—crusts of cobalt on undersea mountains, towering hydrothermal vent spires of copper and gold, and ancient potato-sized polymetallic nodules on the sea floor. As countries and corporations stake their claims, scientists warn that not enough is understood about the risks—dredging up the deep, they argue, could wipe out ecosystems crucial for our planet.

So, what would mining mean for the deep sea—and us? And what's down there anyway?

WHAT'S DOWN IN THE DEEP?

About 30 years after Walsh and Piccard's descent, Australian geologist Ray Binns climbed into a Russian submersible to visit an undersea mountain off Papua New Guinea. 'It was just me, a pilot and an engineer, three kilometres down,' recalls the now retired Binns. 'When we got back to the surface, after about 17 hours, the ship had lost us. It was five miles off course.' Rescue did come, eventually.

What the trio saw under those waves has never left Binns. 'It's completely black,' he recalls. 'Then suddenly the sea floor appears and it's white, like a snowfield. It was like going to the surface of the Moon.' Binns, who studied Moon rocks and discovered Papua New Guinea's hydrothermal vents on a voyage in 1991, actually turned down job offers from NASA. 'I nearly became an astronaut. Instead, the Russians said I became a hydronaut.'

The deep sea begins about 200 metres below the surface. And it really does look alien, says British marine biologist and explorer Jon Copley. 'It's shaped by different forces than the surface.' There are 'immortal' jellyfish, tinier than a fingernail, that change form and are reborn, phoenix-like, over and over; and snailfish that can withstand extreme pressures but at the surface will overheat and fall apart in your hand.

This far from the sun, plants and animals cannot survive by photosynthesis: anglerfish dangle a fleshy 'rod' tipped with a bioluminescent lure in front of their razor-sharp teeth. The world's deepest diving mammals, beaked whales, have bendy cartilage around their ribs that allows them to hunt for giant squid at bone-snapping depths of nearly three kilometres. Some four kilometres down, on average, is the sea floor, also known as the abyss, a sprawl of muddy plains, ragged mountain ranges, canyons and trenches formed by the planet's tectonic plates colliding over millennia.

Some four kilometres down is the sea floor, also known as the abyss, a sprawl of muddy plains, ragged mountain ranges, canyons and trenches.

Entire ecosystems blossom around the hot magma that spews out of cracks in the Earth's crust—shrimps with eyes on their backs and ghostly Yeti crabs clinging to the 'lumpy hills of frozen lava', as Copley describes it, or gushing hydrothermal vents 'like hot springs under the sea'. Washed up on shore, some creatures from the deep really do look like monsters: three-tonne sunfish and giant squid with dinner-plate-sized eyes. The deepest parts of the ocean are the hadal trenches, named after Hades, Greek god of the underworld. The deepest of these is the 11-kilometre-deep Mariana, in the West Pacific Ocean. At these depths, fish, including the yap hadal snailfish, have special proteins to stop the weight of the water crushing their cells, and anti-freeze properties in their blood, so it can flow in the cold.

More than 150 years ago, in 1872, the HMS *Challenger* trawled deep seabeds on a major exploration of the Atlantic, Pacific and Southern oceans, bringing up an exotic collection of rocks, creatures and coral that's still treasured by museums around the world. It offered a small taste of what's down here. 'Every dive looks different,' says Copley, who went on the first journey to the world's deepest hydrothermal vents, in the Caribbean Sea, in 2013. 'It's just as diverse as the mountains and forests and deserts on land.'

WHAT DOES THE DEEP SEA MEAN FOR US?

Biologists estimate we have yet to discover two-thirds of deep-sea species, some of which could help open new frontiers of science. Unlocking the secrets of ancient bacteria and immortal jellyfish, say, could help medicine conquer superbugs, or even reverse ageing.

Copley points to the scaly foot snail, which grows iron plates on its underside in the hydrothermal vents where it lives in the Indian Ocean. 'At spots like this, you can have 400 species found nowhere else on Earth somewhere half the size of Disney World,' says Copley. The snail produces rare nanoparticles that scientists need to produce solar panels. 'But until we saw how the snail made them, we didn't know how to do it ourselves.'

Biologists estimate we have yet to discover two-thirds of deep-sea species.

Life on Earth may have actually begun in the darkness of the deep. At hydrothermal vents of the ancient 'Lost City' in the Atlantic Ocean, scientists are studying the conditions that some speculate gave rise to the first organisms. The ocean makes life at the surface possible—its rich food chain and powerful currents drive the world's weather and support fish stocks. The weight of the sea drags things down; it's the world's biggest carbon sink, compressing dead plants and animals over millennia into fossil fuels. 'The deep sea is the last frontier, but it's not an isolated box,' says ecologist Jeff Drazen at the University of Hawai'i at Mānoa. 'We're still connected to it.'

If you ask some explorers, the key to solving climate change could lie at the bottom of the sea. Many of the rocks that the *Challenger* brought to the surface contained copper, zinc, cobalt and other metals that are now needed for the batteries and wind turbines used in green energy. Manganese nodules (also known as polymetallic nodules) are of particular interest—these small, potato-shaped rocks formed, under intense pressure, on the ocean's abyssal plains over millions of years. There are also crusts of cobalt on sea mountains and towers of copper, gold and rare-earth metals found on hydrothermal vent 'chimneys' where super-hot water hissing from the Earth's crust hits the cold ocean, igniting all kinds of chemical reactions. 'We call them black smokers but they're actually sparkly too, they glitter with crystals of iron, fool's gold,' says Copley.

Worldwide demand for metals is expected to more than double by 2060, according to the Organisation for Economic Cooperation and Development (OECD), and some analysts fear shortfalls could delay the transition to green energy. The Clarion-Clipperton Zone (CCZ), which stretches between Hawaii and Mexico, is so studded with polymetallic nodules that it's estimated to hold more rare metals than all the reserves on land.

The Metals Company (TMC) calls the nodules 'a battery in a rock', arguing they can be mined with less environmental impact under the sea than similar ores on land. It says the haul from its claims in the Pacific Ocean alone could provide metals to manufacture batteries for 280 million electric vehicles. And the world can't afford to wait.

The Clarion-Clipperton Zone, which stretches between Hawaii and Mexico, is estimated to hold more rare metals than all the reserves on land.

Already, global warming is affecting the deep sea—huge volumes of ice that have melted at the poles have slowed critical deepwater currents in the Atlantic Ocean and near Antarctica. A slowdown below could lead to a catastrophic breakdown in the warm waters of the Gulf Stream higher up and fuel wilder weather at the surface. These deep currents also carry oxygen down from the higher levels of the ocean, taking hundreds of years to make the full journey (starting in the Arctic and ending in the CCZ). 'Oxygen levels are already on track to drop by 10 per cent, on average, in the deep,' says Copley, who fears the creeping advance of oxygen 'deserts' down below. 'That's already baked in.'

Some creatures, such as the vampire squid, have evolved to survive in low-oxygen waters. 'But for most it's going to push them into new parts of the ocean. Fierce Humboldt squid will end up in our fisheries as a new predator. It'll throw everything out of whack.'

Of course, deep-sea mining will stir things up itself.

HOW WOULD DEEP-SEA MINING BE REGULATED?

The crew of the MV *Hughes Glomar Explorer* were in high spirits as they sailed from California on a special mission in 1974. 'Mission impossible?' recalled one crew member. 'Nonsense! "Impossible" was not in our vocabulary!' The state-of-the-art ship, named after its billionaire backer Howard Hughes, was on the hunt for rare metals in the deep Pacific.

At least, that's what everyone thought. Decades on, a declassified CIA report revealed the *Glomar*'s true mission: recovering a sunken Soviet submarine and its nuclear-tipped torpedoes. But the cover story that it was a drillship undertaking 'advanced ... deep-ocean mining' was so convincing that even the Soviet Navy, spotting the *Glomar* bobbing in the sea north-west of Hawaii, eventually wished them all the best and left.

Now, half a century on, the world may be on the brink of a real deep-sea mining rush. Robot rovers resembling giant bulldozers have been tested in the deep, trundling over the ocean floor to suck up nodules or carve off mountaintops with metal teeth. A little-known United Nations authority is finalising international regulations that will allow them to start work.

Countries can mine in their own territorial waters (which Japan and Norway, for example, both look to do). But the main game is the high seas, beginning 200 nautical miles from shore, where international law rules. The deep here is simply called 'the Area' by lawyers. Covering just over half the Earth's entire seabed, it's considered the common heritage of mankind, akin to the Moon and outer space.

That means the profits derived from any riches beneath the waves are meant to be 'for the benefit of all', according to the 1982 UN Convention on the Law of the Sea. In 1994 the International Seabed Authority (ISA) was set up to govern deep-sea mining in the Area and divvy up profits fairly among nations. A mining code is still under negotiation, so commercial mining isn't yet allowed. But the ISA sells exploratory licences to countries, and to companies sponsored by governments, to prospect for metal reserves for US$500,000 a pop.

More than 1.5 million square kilometres of international seabed, roughly the size of Mongolia, already fall under these licences, mostly in the CCZ. The zone is also teeming with life. Drazen, who has visited the CCZ many times, describes dumbo octopuses with big, 'ear-like' fins and 'gummy squirrel' sea cucumbers that look like rubber sneakers.

Small Pacific nations such as Nauru and the Cook Islands, which sponsor companies that have licences in the CCZ, argue the billions of dollars earned from mining metals beneath the waves could help lift their countries out of poverty. But others, including Fiji, Palau and Samoa, have called for a ban, worried that mining the deep could imperil the fish and sharks in higher waters.

In 2023, UN member states agreed on a treaty laying the foundations for vast new marine reserves. It was a big win for conservation, nearly two decades in the making, but it did not overrule existing laws for the high seas, such as for shipping; down in the Area, deep-sea mining remains under the ISA's remit. And a growing chorus of countries including Australia, Chile, Germany and Italy (as well as the Vatican) have raised concerns over how the ISA is governing mining prospectors.

One problem, says ISA observer and environmental lawyer Duncan Currie, is that competing interests lie at the heart of the ISA, an agency with more autonomy than other UN bodies. Headquartered in Jamaica, it has a legal responsibility to safeguard deep ocean ecosystems, yet its role is to oversee mining—an inherent tension. Companies and governments that want to cash in on deep-sea minerals would be expected to pay the ISA royalties—a yet-to-be agreed share of profits—for administration costs, and to divide among member states. But there are concerns that under the ISA's proposed models, most of the profits will go to private companies. The ISA has defended its integrity and independence.

Jon Copley notes that, for all the UN's 'utopian vision' of using the sea's riches to correct inequality, the ISA's reliance on private companies to actually do the mining could widen the gap between rich and poor countries. 'Those companies don't work pro bono,' he says. 'Maybe a little benefit trickles down. But a small group are still getting proportionately richer, the opposite of [the original] vision.'

WHAT ARE THE SCIENTIFIC CONCERNS OVER MINING THE DEEP SEA?

Polymetallic nodules in areas such as the CCZ are as crucial to deep-sea ecosystems as trees are to forests, Dr Helen Scales writes in her book *The Brilliant Abyss*. They provide nurseries for octopuses and hiding places for worms. 'When we go down for our studies, about half of the large animals we see depend on them,' says Jeff Drazen, who has conducted baseline ecosystem research funded by TMC. 'And they won't grow back for millions of years.'

Mining the sea floor would dredge up storms of dust and mud that could spread up to 100 kilometres, according to Drazen and other experts. They could smother life and feed toxins into the wider food web, from whale sharks to swordfish. Excess seawater and mud sucked up from the sea floor when the nodules are extracted would be dumped back into the ocean. 'That might add in more fine particles of metal as those nodules are broken up,' Drazen says, including the mercury they contain in abundance. 'Mercury already builds in animals naturally. Remember, your plate of sashimi is only ever one step in the food web removed from [the deep].'

Metallic nodules are also slightly radioactive, and a paper in *Nature* in 2023 raised concern that alpha emitter particles found in them could be released by deep-sea mining. 'They don't penetrate

> Mining the sea floor would dredge up storms of dust and mud that could . . . smother life and feed toxins into the wider food web.

the skin, so they're not normally a problem but they can be toxic when ingested,' explains Copley, who is running United Kingdom–funded research examining the impact of deep-sea mining tests. (Even after 26 years, biodiversity failed to recover at one experimental mining site in the Pacific, for example.) 'It'll be about scale,' he says. 'Could it be managed regionally? Maybe. But we hadn't even considered radiation before now.'

TMC and other prospective miners say they have invested millions of dollars in deep-sea research and have developed technology, such as artificial intelligence and sensors to monitor mining activity, to minimise disruption. TMC's Australian chief executive, Gerard Barron, stresses that many of the concerns about mining remain speculation. 'People first said the sediment plume would find its way to the other side of the world; now we know it's localised,' he says. 'There will always be impact', but compared to the destruction of land mining, he considers deep-sea 'harvesting' to be a game-changer. 'We feel we have enough data' to do it responsibly, he says.

But not everyone agrees. In 2022, more than 30 deep-sea experts, including four who sit on an ISA committee, concluded there wasn't enough data to decide how to regulate mining in the deep. Hundreds of scientists, as well as the governments of Germany, Canada, Spain, Chile and New Zealand, among others, have called for a moratorium on deep-sea mining until further research is done. France, meanwhile, wants an outright ban. Tech giants including Google and Samsung and car companies BMW, Volvo and Volkswagen also back a moratorium.

Other experts say the need for seabed metals has been overhyped. Some electric-battery makers, including Tesla, are already using alternatives to cobalt and nickel. 'And we have these metals on land,' says Currie. 'The real issue is strategic access.' Cobalt and nickel are concentrated in places such as China and Russia, he says, and mining coltan, another metal used in batteries often involves tearing up rainforest or exploiting child labour in the Congo.

Meanwhile, China has the technology to lead the deep-sea gold rush, says marine governance researcher Pradeep Singh, who has spent time on the country's deep-sea vessels. But he doubts it's in China's best interest to kickstart undersea mining when it already controls much of the access to these metals on land. 'If it threatens their dominance, that's when they will jump the queue,' Singh says. Until then, he adds, 'they are happy to wait because whoever goes in first will have lots of technological issues, lots of legal issues.'

Renee Grogan worked for a decade in the Australian mining industry before joining Impossible Metals, a startup that promises to address some of deep-sea mining's environmental risks with clever robotics. 'We're never going to understand all the animals that live in the sediment, so we have to engineer in a way to protect them,' she tells us. Instead of dredging and sucking up nodules, Impossible Metals' prototype robot hovers over the seabed and uses artificial intelligence originally developed for fruit picking to check if there is life on a nodule. Grogan says the prototype would 'harvest' only nodules that appear to be uninhabited.

If Impossible Metals can prove that the technology is effective, Grogan says, it will look to sell it to prospective miners with existing ISA licences. 'It's in no one's interest to jump the gun and start mining before the regulation is ready,' she says. And, while all eyes are on a draft mining code, she warns little attention has been devoted to how it will be enforced. She points to deep-sea fishing—an industry whose strict regulations are poorly enforced due to the sheer difficulty of policing so much ocean. 'We need to talk enforcement now.'

Drazen worries that one of the biggest decisions about our planet's future is being made largely in the dark. We can stare up at the night sky and wonder at the stars, he sighs, 'but we don't think about the deep sea, even though it's our own backyard. The public need to wake up to this.'

While Copley says he has sympathies on both sides of the debate, he keeps thinking about that scaly foot snail. 'If we'd just wiped it out with mining, we'd never have learnt about the solar panels particle,' he says. The snail has become the first species to be listed as endangered because of the impending threat of deep-sea mining.

'The deep is like nature's library,' Copley says. 'We want to read it, not burn it down.'

21

WHAT CAN OUR TEETH REVEAL ABOUT US?

Our teeth are essential—and they carry our secrets long after we're gone. What do they know? And what makes a perfect smile?

Angus Holland

There are good teeth and there are bad teeth. Then there are commercial-grade prosthetic teeth. These are the kind Chris Lyons makes: perfect replica teeth, fantasy teeth, horrible teeth.

From the outskirts of a little market town in Buckinghamshire, England, Lyons runs Fangs F/X, which makes 'character teeth' for film and television. He helped transform Meryl Streep into Margaret Thatcher for the film *The Iron Lady*. He did Rami Malek's teeth for his role as Queen singer Freddie Mercury in *Bohemian Rhapsody*, creating an overbite so severe the actor needed a year to get used to it. He helped turn actor Rufus Sewell into Prince Andrew for the Netflix drama *Scoop*. Then there are the zombies, including for the TV series *The Last of Us*. 'I think we did 180 zombies for *World War Z*,' Lyons tells us, not to mention the undead 'white walkers' in *Game of Thrones*.

If it's decided that a character, such as Meryl Streep's witch in *Into the Woods*, would have had poor dental health, Lyons will make their teeth chipped, crooked and yellow. And it wouldn't have been fitting for one of the stars of *Ten Pound Poms*, a series about Brits coming Down Under in the 1960s, to have perfect teeth. 'She'd had her teeth done, and it was just a bit too nice for the period, so we had to break them up a little bit,' he says.

> Our teeth can ... be used to identify us even if, literally, nothing else remains. They can speak to our social standing and income level.

As Lyons knows, our teeth can say a lot about us. They can be used to identify us even if, literally, nothing else remains. They can speak to our social standing and income level. Cosmetic procedures, such as veneers and teeth whitening, can reveal our anxieties. And we, in turn, can make statements with our teeth. The

Vikings filed grooves into theirs, today's rappers have gold replacements and diamond implants. Shimmering 'grillz' and sparkling crystal tooth gems are all over TikTok.

Even if we're unlikely to die from a cavity these days, life without teeth would be pretty awful. As 16th-century Spanish author Miguel de Cervantes observed, 'Every tooth in a man's head is more valuable than a diamond'.

What's so unusual about teeth? How should we look after them? And can a bite mark really send somebody to jail?

HOW DO TEETH WORK?

Most animals use their teeth in specific ways. Sharks use them to slice and dice prey (they don't chew). Cats' razor-sharp twin canines grasp and tear and their incisors help with grooming. Dogs rely on canines to shred, incisors to grab, and molars to chew and crush. Adult cats have 30 teeth, dogs 42. Giant armadillos in South America's rainforests have the most teeth of any mammal, 74 peg-like stumps for crunching insects. Great white sharks have slightly fewer in everyday service—around 50—but more in back-up rows, adding up to several hundred. Depending on your definition of 'teeth', it's thought sea slugs have the most—several thousand.

Thanks to our omnivorous diet, a set of human teeth is more like a Swiss Army knife. 'Incisors help us bite into fruit, pointed canines and their moderately crested posterior neighbours pierce and tear meat and plants, and low-crowned molars help us grind up hard foods, such as nuts and seeds,' writes Tanya Smith, a professor at the Australian Research Centre for Human Evolution at Griffith University in Queensland, in her book *The Tales Teeth Tell*. Our teeth, she says, magnificently balance efficiency and over-specialisation. In the past, they had to: without good teeth, we were goners. 'If your teeth didn't fit together, or they didn't come out in time, or they didn't have the right shapes, you might not make it to adulthood,' Smith tells us.

Our teeth start developing under the gum line when we're in the womb and begin to erupt, much to our discomfort, when we're about six months old. The incisors emerge first and molars last, 20 teeth in total. After seven to 10 years, they're replaced by a permanent set of 32 teeth (cue the tooth fairy, turning a loss into the gain of a strong new tooth). Wisdom teeth, four molars at the very back of the jaw, are the last to appear, often in our late teens or early twenties.

Each tooth has a bundle of blood vessels and nerves in its core, known as pulp, where teeth 'feel' things, including toothache when you get a cavity. Long roots anchor the teeth in the jaw bone and draw in blood. The pulp is covered with hard yellow dentine, which in turn is covered in enamel, sometimes several millimetres thick, to create the crown.

Our teeth are not that different from those of our prehistoric ancestors, which can be a problem in an era where we cook, mince and julienne our food. These days, 'they don't fit quite right,' says Smith. 'They get stuck. They still think we're living in the Stone Age.' Impacted wisdom teeth, which don't emerge properly, are possibly casualties of our teeth not all fitting in our jaws, which have slowly shrunk over time through lack of hard use. Happily, we have also evolved to invent dentistry.

WHAT CAN OUR TEETH REVEAL ABOUT OUR PASTS?

For anthropological researchers such as Tanya Smith, teeth are a rich source of information. Not only can they survive for aeons relatively intact (teeth between 63,000 and 73,000 years old have been found in South-East Asia), but their chemical makeup and wear patterns can reveal much about their owner—what they ate, where they came from, even the season of their birth.

'You can paint this really detailed picture of the biology of an ancient human or even a living person today,' says Smith. 'I could tell, potentially, how long you nursed when you were a kid, how often you were sick, whether you moved from one region to another. I might be able to tell what season you were born in.' Even a particularly stressful day—such as the one on which you were born—leaves a line on your teeth that is visible under a microscope.

A particularly stressful day—such as the one on which you were born—leaves a line on your teeth that is visible under a microscope.

Judith Littleton, a professor in anthropology at the University of Auckland, studied prehistoric teeth found in Bahrain. 'Cells stop forming enamel because the body is stressed, and when they start again, it leaves a line,' she says. 'You can track the secrets of those lines through childhood, a history of a child's exposure to stress and their habits over time.' Littleton also examined the differences in diet between Bronze, Iron and Islamic-age people, noting variations in the way they processed food, which caused their teeth to wear differently. She suggests that a prime cause of prehistoric cavities was dates—sugary and sticky, they may have introduced *Homo sapiens* to wholesale toothache.

Other clues are very particular. When researchers inspected the teeth of a woman buried at a German monastery around the 11th century, they made an odd discovery: more than 100 particles of ultramarine, a pigment that comes from the precious stone lapis lazuli, encased in her plaque. What was it doing on her teeth? Several scenarios were considered, including that the lapis lazuli had a medicinal use—it was used to treat scorpion bites and ulcers in first-century Greece—or that the woman had been kissing religious images (it was not unheard of in Europe in the 14th century, but that was long after her time). In the end, the researchers, from the University of York and the Max Planck Institute for the Science of Human History, thought it most likely the woman had been producing manuscripts, and had licked her paintbrush from time to time to smooth it into a fine point (a known practice). The monastery's manuscripts had been destroyed in a fire long ago but the woman's teeth added to evidence that monks were not the only ones making books in the Middle Ages.

Rita Hardiman, associate professor at the Melbourne Dental School, is studying thousands of teeth dusted off during excavations in Melbourne to learn more about the health of our Victorian-era ancestors. The source of the extracted teeth: dentists who threw them down the drain. 'We're hoping to learn about the population in terms of who was going to see the dentist and why,' says Hardiman. 'I suspect a lot of the "why" is because things were getting pretty dire, and they couldn't wait any longer.'

Some of the teeth were worn down by the use of pipe stems while others appeared to be healthy. 'If you've lost numerous teeth to decay or gum disease, it might have been more convenient to take [the rest] out to place a full set of dentures,' says Hardiman. Indeed, earlier last century, young people sometimes had their teeth extracted and replaced by dentures to save the trouble later. The procedure could be a 21st birthday gift or a step in enlisting in the army. During World War II, the Australian Army Dental Corps pulled more than 1.5 million teeth and made around half a million dentures. Or, says Jacqueline Healy, senior curator at the University of Melbourne's Henry Forman Atkinson Dental Museum, 'If you were getting married, what your parents might do is have all your teeth extracted and have you fitted with dentures so you wouldn't be a burden to your husband.'

Earlier last century, young people sometimes had their teeth extracted and replaced by dentures to save the trouble later.

Today's dentists can glean a lot just from peering into your mouth. Dr Andrew Gikas, a Melbourne dentist who specialises in bruxism, or teeth grinding, might be able to tell if you grew up in the city or countryside, for instance. 'If you have a whole bunch of fillings, you probably grew up in a non-fluoridated area,' he says. 'That's my first guess.' (The first place to add fluoride to water was Beaconsfield, Tasmania, in 1953 but take-up was patchy and controversial. Sydney's water wasn't fluoridated until 1968, Melbourne's in 1977.) 'I can probably tell how aggressive you are into brushing by how much you've pushed your gums down,' Gikas says. 'I can tell whether you're into acidic drinks—Coca-Cola and soft drinks—because I'll see some erosion on your back teeth.'

HOW CAN TEETH BE USED TO IDENTIFY PEOPLE?

Former dentist Richard Bassed, who runs the forensic odontologist department of the Victorian Institute of Forensic Medicine, finds teeth extremely useful. Made of enamel, the hardest substance in the human body, they can be all that remains of a person if they perished in a fire or decomposed beyond recognition. Even a single tooth can be used to make a positive match, says Bassed, who helped identify victims of the Black Saturday bushfires in 2009, the Bali bombing in 2002 and the 2004 Boxing Day tsunami in Asia.

Sitting in his office, he pulls up a photograph on his screen showing the aftermath of a bushfire. It's mostly just a pile of grey ash. 'You can see there's almost nothing left,' he says. 'You can see why police officers searching the scene aren't going to find the teeth. We go there with sieves and fossick around in the ashes.' Then there's a hunt for the victims' dental records. If families can't help, calls might go out to dentists in their area. 'Fragments of roots, fragments of crowns—you put those all in the right anatomical position and see if there's any antemortem [pre-deceased] records that will match,' says Bassed.

Finding the teeth and teeth fragments and putting them together can be a time-consuming process, 'but it's rewarding because that might be the only remains of that person that their family will get back. It's either that or nothing. And at least they get the certainty of knowing, yes, this person died, and they're not holding out hope.'

In criminal cases, bite marks have been largely debunked as a means of identifying people, says Bassed. Flesh moves around, making bite marks inexact. (You might be able to differentiate between a bite made by an adult or a child, he says, but that's about it.)

In criminal cases, bite marks have been largely debunked as a means of identifying people. Teeth can leave clues in chewing gum or chocolate, though. 'It's a proper impression of your teeth that isn't subject to all the vagaries of human skin and healing and movement and all that sort of stuff,' Bassed explains. Indeed, a burglar who took a bite out of a hunk of cheese in a Los Angeles house in 1985 went to jail for four years after a dental examiner identified him using an unusual groove in the remnant.

WHY DO WE VALUE STRAIGHT, WHITE TEETH SO MUCH?

Australian soprano Dame Joan Sutherland may have been blessed with a magnificent voice, but the gods were less generous with her teeth. 'You can't sing like that any more. We'll have to do something about it,' her husband, conductor Richard Bonynge, told her, according to a 1963 article in *The Australian Women's Weekly*. She had her teeth capped at enormous expense for a 'Covent Garden soprano whose £30 a week salary had to support a household of five and pay the rent'. (Sutherland later had to have them recapped after an infection.)

La Stupenda's tooth troubles were not unusual. For centuries, the consumption of refined sugar has devastated dental health, feeding bacteria—chiefly *Streptococcus mutans*—that live on our teeth. The bacteria then produce lactic acid, which erodes enamel, causing decay. 'People think rich people had good teeth and poor people didn't,' says Jacqueline Healy. 'In fact, it was the opposite—because what did rich people have? Sugar.'

In 16th-century England, aristocratic women blackened their teeth to emulate Queen Elizabeth I, who had a sweet tooth. In the early 1800s, Napoleon Bonaparte was fastidious about tooth care and gave his second wife, Marie Louise, a box of gold-covered dental instruments.

While there is evidence of dentistry going back to the Stone Age, by the 14th

'Part of the entertainment of going to the market would be watching people get their teeth taken out, without any anaesthetic or antiseptic.'

century in the West, 'dentists' were considered to be for the brave or desperate; their toolkits included the pelican, which resembled the seabird's bill, used to grip and lever out teeth, invariably damaging other teeth and gums in its wake. 'Part of the entertainment of going to the market would be watching people get their teeth taken out, without any anaesthetic or antiseptic,' says Healy. 'Often, a bit of your jaw might come off.'

The modern profession is generally traced back to French physician Pierre Fauchard, who in 1728 published the first dental textbook, *The Surgeon Dentist*. One of his innovations was improving the appearance of dentures by enamelling them with the assistance of potters. Yet many techniques around this time remained crude. Early false teeth had sprung hinges and a proclivity to leap out of the wearer's mouth. Because ivory and porcelain dentures were so expensive, cheaper 'Waterloo Teeth' were made from real teeth looted from battlefields, morgues and graveyards.

Tooth decay continued its march into the 20th century. 'The most obvious sign of undernourishment is the badness of everybody's teeth,' observed George Orwell in his 1937 treatise on life in northern England, *The Road to Wigan Pier*. One woman told him bluntly: 'Teeth is just a misery.' Even today, the problem is rife worldwide: a recent study suggested more than a third of people have untreated cavities or decay.

It makes sense, then, that people aspire to having straight white teeth. Yet as late as the 1970s, cosmetic dental work was far from routine, even for celebrities. David Bowie straightened and whitened his typically 'English' teeth only after he married the supermodel Iman in 1992. Having your teeth 'done' was still seen, at least in Britain, as vain. In 1995, novelist Martin Amis was roasted in the press for reportedly spending nearly $40,000 getting his (chronically troublesome) teeth pulled and replaced with implants. Who did he think he was!

That same year, though, *The New York Times* reported that recession-hit middle-class American families were prepared to cut back on anything—except having their children's teeth straightened, which was seen as a vital social leg-up. 'If you go into a job with teeth out of a novelty store, people aren't supposed to discriminate,' one orthodontist said. 'But people do.'

Seemingly perfect teeth reinforce class differences as symbols of social advantage, argue Abeer Khalid and Carlos Quiñonez from the faculty of dentistry at the University of Toronto. 'A distinguishing feature of North American society is preoccupation with self-image,' they wrote in the journal *Sociology of Health and Illness* in 2015. 'Nowhere is this more apparent than in the prevailing fixation with straight, white teeth.'

In 2007, a paper published in the *British Dental Journal* stated: 'Subjects with a relatively normal dental appearance are judged as better looking, more desirable as friends, more intelligent and less likely to behave aggressively, and teachers have higher expectations of them.'

While some dental treatments have a clear functional purpose—heading off the long-term ill effects of misaligned jaws, uneven bites and cavities, all of which can cause issues elsewhere in the body—others are purely cosmetic. Veneers, for example, are thin shells made of dental composite or porcelain that fit over real teeth to give the appearance of a dazzling smile. First used in the 1930s, when they were crude and stained easily, today they come in various shades of white to suit different skin tones. 'We've definitely seen a shift from having dentures and anything removable,' says Peter Laird, a dentist at Glenferrie Dental in Melbourne. 'People are wanting something a lot more immediate, a lot more permanent, and not having to worry about teeth moving around anymore.'

Subtlety is in vogue, too. Catherine, the Princess of Wales, apparently realigned her teeth with a procedure called micro-rotation, in which her outside front teeth (incisors) were moved slightly with braces on the backs of her teeth. The result was a slightly imperfect, more 'natural' smile. In Japan, where there has long been a tradition of finding beauty in imperfection, called *wabi-sabi*, slightly immature-looking canines called *yaeba*, or 'double teeth', are coveted by some young women.

Is whitening, or bleaching, your teeth safe? Yes, says Dr Scott Davis, president of the Australian Dental Association, but only if it is done under dental supervision, which can pick up potential issues such as cracked teeth and fillings that are another colour. 'Hopefully people are looking after their teeth not just because it makes them look good but because it's good for their overall health,' he adds. 'White teeth don't necessarily mean they are healthy and it isn't natural for older people to have bright white teeth.'

As for general dental health, only 31 per cent of Australians have regular check-ups, says Davis. They aren't covered by Medicare, although 'there are more reasons why people do or don't see a dentist regularly than just money—there are complex behavioural issues as well.' Still, he notes, check-ups can save expensive fixes down the track. Flossing is one simple favour you can do your teeth every day: it removes about 40 per cent of food remnants that cause plaque, while brushing twice daily removes the rest (but not too soon after you've consumed anything sugary because sugar can temporarily weaken enamel).

Meanwhile, dental embellishments are having a moment. 'Grillz' and tooth gems are usually removable adornments that add a flash of bling to a smile. Both are throwbacks to ancient times: Etruscans used gold and jewels to enhance teeth and Vikings filed theirs for effect—as if Vikings weren't already scary enough.

Ashlin Carlisle learned about affixing gems to teeth on a trip to London and now runs a small business called Pearlyygems from her Sydney home. She mostly sells crystals, which are non-toxic in case you accidentally swallow one. Carlisle teaches her clients how to look after their gems and advises them to visit a dentist when they want them removed. How long they last, she says, can vary. 'It's all dependent on the size and the placement of the gem and how the client follows their aftercare.'

In Perth, dentist Maheer Shah—aka 'Dr Grillz'—has expanded into decorative grillz popularised by US rap artists such as Post Malone, who reportedly spent more than US$2 million ($3 million) on 18 porcelain veneers, eight platinum crowns and two six-carat diamonds to replace his upper canines. Shah's grillz are less intrusive: made of gold or silver, they sit over the front teeth like a custom-made mouthguard. 'It's becoming so much more accepted, like tattoos,' he says. Occasionally, a client will want something more permanent, which he advises against. 'It's pretty rare, but one guy bought eight gold teeth,' says Shah.

After the success of *Bohemian Rhapsody*, Chris Lyons made a set of gold Freddie Mercury teeth for actor Rami Malek to wear to events. Mostly, though, Lyons works with normal dental composites to fashion his prosthetic gnashers,

> 'Grillz' and tooth gems add a flash of bling to a smile. Both are throwbacks to ancient times: Etruscans used gold and jewels to enhance teeth.

such as those for Tilda Swinton in the film *Snowpiercer* or for Sewell in *Scoop*. 'Prince Andrew's teeth are quite particular,' he tells us. 'The more subtle stuff is actually more challenging than the big monster stuff. You want people to look at it and think, *Have they done something?*'

22

HOW DOES A PERSON GET LOST AND FOUND?

It took her just moments to lose sight of the track in the wilderness.
Days later, as searchers combed heavy forest, time was running out.

Jackson Graham

Alll it took was a wrong turn. Madeleine Nowak was the straggler on a day hike with her partner and friends when she came to a huge fallen tree blocking her track. She tried to scramble over it but couldn't. She tried to go around it—just a quick detour—but when she got to the other side, the track was hidden under leaves. Where had it gone? The path must be nearby. She trudged a bit. A bit further. Suddenly, she was deep in the forest.

'Cooee!' she called.

Silence.

'Cooee!'

Nothing.

Madeleine, then 73, had always had a lousy sense of direction. An amateur photographer, she'd amble behind her companions on hikes, snapping birds and plants. Her friends and her partner, Clive, knew she was a dawdler; Clive was used to backtracking to find her on walks. On this afternoon on the Queensland island of K'gari, formerly Fraser Island, she was convinced he'd hear her calls. When nobody came, she kept walking. 'I guess I followed my nose,' she recalls.

Afternoon turned to night and night turned to day, and there was still no sign of anyone else. By now, emergency services were combing tracts of dense ferns and towering satinay trees for Madeleine. But she was on the move, bashing through the bush in search of the coast.

Emergency services were combing tracts of dense ferns and towering trees ... But she was on the move.

Around 38,000 people go missing in Australia every year. There are many reasons: escaping domestic violence, disorientation brought on by health conditions, simply failing to tell anyone where they are. In about 3000 cases a year, search and rescue operations are required. About two-thirds of them are at sea; land searches might involve people who are injured, bogged and out of reach of help or lost. They're often found swiftly but occasionally they're not—and every passing hour reduces their chances of survival.

How do people get lost? What should you do if you lose your way? And what will searchers do to find you?

WHO GETS LOST?

When following her nose failed, Madeleine tried to correct her course. She walked uphill to try for phone reception or to spot a landmark, but at the top, there was no signal, and the rainforest was thick in all directions.

Oh dammit, now what have I done? she recalls thinking. *How am I going to sort this?*

'Once you leave a track, unless it's really well worn, they're very hard to see from a few metres away,' says Jim Whitehead, who was involved in the search for Madeleine. He's no stranger to the bush. 'I used to bushwalk in my younger days,' he says. 'I'd never tell anybody where I was. I did all the things I tell people not to do now.' Whitehead wrote the latest version of the *National Search & Rescue Manual* and estimates that, as state search and rescue co-ordinator for Queensland Police for 15 years until 2022, he was involved in about 24,000 search and rescues, or SARs.

When he started in the force in the 1980s, 'people said there would [one day] be no "S" in SAR,' he tells us. 'It would just be a rescue because we'll know exactly where they are. Forty years on, yes, we do have GPS as we have locating beacons, we have a tonne of electronics. But basically, it still comes down to someone searching for a missing person. We haven't actually got to the stage yet where we can just find somebody by hitting a button . . . It's still very manual.'

Typically, it's novice bushwalkers who get lost in the bush, he says. 'They've got no map, they've left at midday in a pair of thongs and shorts, and [they don't realise they] are never going to make the return journey before the sun sets.' Rod Costigan, who leads the volunteer group Bush and Search Rescue Victoria, agrees. 'It will not be the walk leader or the guy who said, *"Come camping with me"*,' he explains. 'The person who is lost alone overnight will usually be someone who came along but got separated.'

Dependents are especially likely to stray. 'For the adventuring types, there are plenty of sources of advice as to how to be prepared,' Costigan says. 'But if you are taking kids for a picnic in a country picnic ground or taking your elderly parent for a short walk near your bush block, you won't be thinking in these terms.'

Hikers who rely on unofficial maps and trails are also prone to losing their way. In the Blue Mountains, west of Sydney, for example, these tracks are not maintained and can be difficult to follow. Most of them have signs warning people of the risk of getting lost. 'Unfortunately, those signs don't always stop people,' says Sergeant Dallas Atkinson, who leads the Blue Mountains Police Rescue unit. A 2023 survey of searches in the Australian bush found that men were more likely to leave a path when they become lost while older females were more likely to be found closer to one. (Madeleine was an exception. 'I kept going,' she says, 'I'm not afraid of being on my own.')

Cars can get bogged out of mobile phone range. 'No one knows about you unless someone comes across your car,' says Whitehead. In May 2023, Melbourne woman Lillian Ip became bogged in Victoria's High Country. There was no phone reception and she was unable to walk for help. After five days, police spotted her from the air. Ip, a teetotaller, had no water but survived on lollies and sips from a bottle of wine she'd planned to give to her mother. (An expert we spoke with said alcohol was not the preferred drink to prevent dehydration but it could extend survival time.)

GPS navigation can send drivers down unsafe roads, such as the two German tourists whose four-wheel drive got stuck in Queensland's far north in February 2024. They trekked 60 kilometres past a flooded river, and through bush inhabited by crocodiles and snakes before emerging a week later in Coen, population 320.

In about a quarter of searches, the lost person has dementia, says Whitehead. One woman with dementia hid in cane fields north of Mackay, Queensland, for three days in 2020. 'We couldn't find her by thermal imaging because she was covered in mud,' Whitehead recalls. 'We couldn't find her by searching because she would deliberately hide.' Luckily, a truck driver spotted her on the side of the road and alerted police, who shifted their search. 'It was only because we saw the cane bend down from the air that we found her.'

Looking for someone on the ocean has its own complications. From the air, a person is a fleck on the surface as currents and wind carry them from their last-known location, says Rick Allen of the Australian Maritime Safety Authority. 'The ocean is often very featureless so it's an intensive activity to try and look for targets,' he says.

HOW DO WE KNOW IF WE'RE LOST?

As a teenager, Madeleine would lose her way on shopping trips. *Where on Earth am I?* she would think. *Which door did I come out of?* On holiday in Athens in her twenties, she misread a map and ended up an hour from where she wanted to be. 'I stopped and asked someone, and they pointed me in the right direction.'

Humans draw on several brain regions, particularly those involved in spatial memory, such as the hippocampus and entorhinal cortex, to build mental maps. 'It's the ability to have a sense of the surrounding environment as a two-dimensional representation,' says Don Montello, a researcher of lost person behaviour at the University of California, Santa Barbara. 'That's what allows creative navigation [and] shortcutting and is tied to maintaining orientation.' A mental snapshot of your surroundings will include landmarks. When travelling, for instance, 'I want to know which direction the front of my hotel is facing,' Montello says. 'Even large macro features like, "Where's the Seine in Paris?".'

As with any skill, navigating depends on innate ability— but practice helps. Before London taxi drivers receive their licence to drive a black cab, they sit a test called 'the Knowledge', proving they've memorised streets, landmarks and hundreds of routes around the city. One study of 16 drivers found they had larger hippocampuses than people who didn't drive taxis. The effect was even more pronounced among drivers who had been in the job longer.

Being lost means being in a state of uncertainty, says Montello. 'There're degrees: I can have no idea or I can think I know [where I am] but not really be sure.' Whitehead offers a simple test. 'When you can't actually point on a map to see where you are going, you're lost.' Still, he says, most people don't want to admit it. 'We all want to appear to people that we know what we're doing.'

Madeleine's partner, Clive, has a good sense of direction. 'Because I've basically relied on him for 45 years, I'm worse,' she says. She can't point out which way north is, for example, although she's not at a complete loss. 'If I can see the sun come up, then I know that's east.'

WHAT SHOULD (AND SHOULDN'T) YOU DO IF YOU'RE LOST IN THE BUSH?

Clive took what he describes as a 'cold' approach after he realised Madeleine was missing from the K'gari track. As a scientist, his way of dealing with things, he says, is not to get caught up in emotions. He couldn't call triple zero without phone reception so he backtracked, yelling Madeleine's name. Two friends on the hike raced to find the driver who earlier that day had dropped the group at the track. They picked up Clive and he called the police. 'There was anxiety, but it was more a case of, "What are the best things we can do to get this sorted out?",' he says.

As sunset approached, Madeleine found a hollowed-out tree to spend the night in. She rummaged through what she had: walking poles, a raincoat, a pear, boiled eggs, rice crackers and cashews. She had enough water to last two-and-a-half days. 'I started rationing,' she says. K'gari is home to about 200 dingoes that roam in packs, so she placed her backpack 20 metres away—its scent wouldn't draw animals to her.

As with any skill, navigating depends on innate ability— but practice helps.

One item Madeleine didn't have was a personal locator beacon. When activated, the matchbox-size device sends a satellite signal to stations across Australia. Hikers register their beacon before setting off, listing contact details for both the hiker and someone who knows their plans and can share trip details with searchers. 'The best thing about them is they are not reliant on a telephone or internet connection,' Whitehead says. 'It's probably the best tool you'll ever buy.'

Phones can create a false sense of security. Keith Muller, a deputy controller in Victoria's State Emergency Service, works on searches in the Werribee and Lerderderg Gorge areas, where he says most walkers assume they'll have mobile phone reception. 'Both the Werribee and the Lerdy, believe it or not, there's hardly any emergency radio reception, let alone phone reception,' Muller tells us. A phone 'chews up the battery really quick', he adds, as it searches for communications towers.

People who get lost in the Blue Mountains, on the other hand, can usually call the police. Atkinson says his team will explain how they can pinpoint their longitude and latitude using their phone. 'We can determine exactly where they are and send a team in to get them,' he says.

Madeleine's phone momentarily worked on the first night. She texted Clive:

> Obviously lost. Curled up in a tree stump for the
> night. Almost out of battery. Will head east in
> the morning looking for tracks or the sea. SORRY.

Then her battery went flat.

Madeleine didn't panic but she was irritated with herself for worrying Clive and her friends. *I've stuffed everybody's holiday*, she thought. Determined to fix things, she lined up trees with the sunrise to determine which way was east and off she went into waist-deep scrub. Her walking poles were all that stopped her from falling over. *I just need to keep myself safe*, she kept thinking.

She now realises that trying to find her own way out was a mistake. Every step took her further from the searchers.

Whitehead says people who are lost should find shelter and 'do nothing until you actually hear searchers come. If you're lost, you're lost. Don't make it harder by trying to un-lose yourself.'

Madeleine could hear helicopters looking for her. She found a clearing and waved her poles in the air. 'They weren't close enough to me. In fact, they didn't go over me,' she says. If she'd known, she might have used her phone screen to reflect light from the sun towards the helicopter. (The lenses of spectacles, reflective blankets or surfaces on vehicles can also work; police night vision can pick up light from torches or phones.) Today, she carries a battery pack to keep her phone charged when she goes walking and knows to switch off her phone if there's no reception, to conserve power.

Sometimes, lost people write words or some sort of signal with rocks or branches in an open area. In 2020, three men were found marooned on a Micronesian island after they wrote 'SOS' in the sand. 'To do something like that, it needs to be in letters about 10 metres long because little letters don't show up from the sky,' says Whitehead.

One of the biggest factors in a lost person's survival is access to water. People can generally go weeks without food, says Dr Paul Luckin, a medical adviser to search and rescue teams across Australia. 'But to survive on no fluid at all, you are generally looking at about three to four days,' he says. 'If you're in an arid environment in the West Australian desert, on the worst of the hot days, you might need to be drinking towards a litre an hour.'

Luckin refutes the myth that people can drink their own urine. The body's

'To survive on no fluid at all, you are generally looking at about three to four days.'

waste will only make dehydration worse, he says. 'Drinking urine is a negative gain.' Seawater has a similar effect, and the salt also affects brain function. 'It's characteristic of people in the lifeboat who eventually drink seawater that they're described as going mad.'

Madeleine, a dietitian, knew how important water was. 'I had frequent, very small drinks; just a mouthful to keep going,' she recalls. She even licked leaves from branches on the third day to get extra fluid. 'I thought, *Now, there's a risk of infection here, however, I reckon I'll be out before it manifests.*'

HOW DO SEARCHERS FIND A LOST PERSON?

By Madeleine's third night in the bush, Clive was increasingly worried. He thought she should already have reached the coast. 'I was quite depressed, quite sad, quite tearful,' he says. 'Friends were plying me with gin and tonics and trying to keep me happy, but then you're on your own. You're sleeping in a bed, and the bed is nice and warm, and your mate, your partner, is somewhere out in the bush.' He offered to help search. 'Their reaction was, "We've lost one person, [we] don't need any more",' he says.

About 60 people were looking for Madeleine. Whitehead didn't expect her to be walking for so long. 'The 70-year-olds that I know don't do that.'

Searchers generally start from the lost person's last-known location: a car parked at the entrance of a national park or a sighting along a trail. Telling someone where you are going is crucial. If you haven't, says Caro Ryan, a search commander with the New South Wales SES, '[we] are starting with such a massive landscape area'. Police draw a circle around the point on a map to show how far the person could have walked. 'If everything is perfect, you have to be inside that circle,' Whitehead says. Generally, people walk downhill. 'Most people will get fatigued and won't try to fight the environment.' Police also note places where they could become stuck and where there are water, fences or roads they might follow.

Searches at sea follow a similar plan: rescue crews will 'box up' an area where they believe a lost person might be. 'Sometimes we'll have a last-known position, sometimes we'll only have a route,' says Rick Allen. Looking at currents and weather, his team will predict how far a lost vessel or person in a lifejacket could have drifted. Sometimes, as many as 12 aircraft search at once.

On land, rescuers might follow the path the person was meant to be on or use a helicopter or drone to scout for them. 'We've had a number of operations resolved very quickly by launching a drone and being able to spot the person from the air, get an exact position of where they are and saying, "We know where you are, we'll come and get you",' says Dallas Atkinson.

If police find no trace of the person, they call for backup—volunteer groups help scour off-track and detectives explore possible scenarios, checking traffic cameras, dashcam footage and security cameras and looking for shoe prints. Sometimes dogs are used, particularly when looking for people with dementia who can wander in any direction.

In dense bush, searchers work in teams spaced out in a row, periodically stopping to listen and calling out the lost person's name. 'We've done tests in our unit of the female voice versus the male voice,' Ryan says. 'Usually, we would use the female voice. The high voice seems to carry further.'

HOW DO AUTHORITIES KNOW WHEN TO CALL OFF A SEARCH?

Paul Luckin first helped to rescue people when he was a paramedic in Hobart, often helping tourists stuck on cliffs along the Derwent River. Later, when he was training as a specialist in anaesthesia in South Africa, his superiors tapped him for a role in mountain rescue.

Near the hospital in Durban where Luckin worked, police would sometimes clear the highway so a helicopter could land and fly him and his team to peaks as high as 10,000 feet in the Drakensberg Ranges, to retrieve lost or stuck mountaineers or victims of plane crashes. He has also worked as an anaesthetist in the Australian Navy, treating war casualties in Afghanistan.

Today, his job is to estimate how long someone would be expected to survive in any given search and rescue case. 'When [the police] phone me and say, "We're up to our backsides in the snow", I understand because I have been in exactly that circumstance,' Luckin says. 'When they phone me and say they are in an arid environment, I understand because I've been in exactly that circumstance in Afghanistan [and] training in Kuwait where it's 54 degrees.'

Chances of survival are largely based on the conditions, the person's physical and mental state, and their access to water and shelter. In Madeleine's case, Luckin estimated water could limit her chances. 'There was no surface water in the area; a sandy base, so there was no possibility of water collecting.' He was told she had only half a litre of water, so he estimated she would be severely dehydrated about 76 hours after she was last seen. 'I thought the probable end of her timeframe for survival was the end of the day on Sunday,' he recalls.

When Luckin concludes that someone's chances are slim, he tells police: 'I look forward to the phone call telling me I'm absolutely wrong, that you've found them alive and well.'

Police will work past the deadline and renew a search if new information comes to light, says Whitehead. 'It's never determined by cost.' In his time, about 2 per cent of people were found dead and around 1 per cent were never found at all. 'We have an obligation to find people,' he says. 'We generally go on for another two or three days to allow for us to find something, a deceased person, a bone, or actually find them if they survived.'

It can be difficult work for volunteers who are called in after the obvious places have been searched. 'We don't have a lot of happy endings,' says Ryan of searches she's been on. Some moments she still treasures: the search for toddler Anthony Elfalak was in its fourth day in the Hunter Region of New South Wales in 2021 when the child was spotted from a helicopter, sitting in a creek bed less than half a kilometre from the family home. Ryan heard the boy's mother as police shared news that he was safe. 'It wasn't a call, it wasn't a yell, but something in between. That's the sound I will never forget. That's the one we rarely get.'

Luckin will never forget the 72-year-old prospector who was lost in blistering heat in the Western Australian outback in 2023. He was eventually found, severely dehydrated, within half an hour of when Luckin thought he would die. 'He survived to the absolute limit. He said afterwards, when he lay down where they found him, he knew he wouldn't be getting up.'

Madeleine had two sips of water left on Sunday morning when she emerged from the forest near where some holiday-makers were camping on the coast at K'gari. 'I thought,

Oh my God, I must look like some bedraggled dangerous creature, so I started by just saying, "Hello, I'm . . ." And they looked at me and said, "We know who you are, thank goodness you're here. What can we do?".' She asked for water, and they called triple zero.

Clive had just wished a group of SES volunteers good luck for the day when he got word Madeleine had been found. But it wasn't until he saw her get out of the police car that he felt a wave of relief and then joy.

'I was the first one to grab her,' he says. 'I was checking her out; she looked a bit gaunt in the face and had a big, embarrassed smile. But she looked fine.'

Madeleine felt relief because Clive and her friends knew she was okay. 'He didn't let me out of his sight for the rest of the day,' she says. 'The police said to me, "You know, it's very rare that at this stage it's a good ending. Three nights is a bit long."'

As a helicopter whisked her to a hospital for tests, she looked down at the island. 'I realised why they couldn't find me. It was just total cover. There's nothing you could see down there.'

She now understands what a wrong step can do. 'That's the whole of life, isn't it?' she says. 'If you make a mistake, it has lots of consequences. And you then have to deal with them.'

23

WHY DO WE NEED A NEW WORD FOR RETIREMENT?

The decision to stop work is the start of a whole new adventure. Is there a right time? And what should you expect?

Angus Holland

For some people, the moment just arrives. 'I'm done with all this' is how Judith Venables remembers her decision to retire from a productive career that included, at one stage, being a zookeeper. At 61, she was suddenly ready to leave her role in children's welfare. 'I just jumped ship.'

> Deciding to retire can be agonising, raising curly questions. What will I do all day? Will I lose my identity and sense of purpose?

For many of us, though, it's not so easy. Deciding to retire can be agonising, raising curly questions. What will I do all day? Will I lose my identity and sense of purpose when I'm no longer the chief executive of a bank, a school principal or a newspaper reporter?

Indeed, while retirement might sound great when you're in your thirties and forties (and there's an entire movement dedicated to retiring as early as possible), crossing that threshold later in life can be challenging. It's not just about quitting your job—it's about entering a whole new stage of life.

So how do you prepare for retirement, if you can at all? When's the right time to go? And how much money will you need?

WHEN IS THE RIGHT TIME TO RETIRE?

'It just got to the stage where I thought it was time to do other things,' says Lawrie Mann from Armadale in Melbourne, who left corporate life when he was 66. 'I'd also had some friends who had died while they were still working. One bloke in particular would have been so unhappy that he couldn't do a lot of the things that he'd wanted to do after he retired because he didn't ever retire.'

For one retired arts administration executive now in his late sixties, the COVID-19 pandemic was the deal-breaker. He liked his job but hated working from home during lockdowns. 'I just got sick of it,' he says. 'And I had the money and I wanted to do things that I wanted to do.'

Their experiences are typical in Australia, where we tend to peg the 'right' time to retire on the age at which we qualify for a federal age pension—currently 67 for those born on or after 1 January 1957. As people live longer and can work for longer, many countries (including Britain) are raising the pension qualification age and not everyone is on board.

The French government's decision to gradually increase the country's retirement age from 62 to 64 saw protesters take to the streets in 2023, holding up signs reading 'Retirement before arthritis' and 'Leave us time to live before we die' (they were probably a little snappier in their native tongue). President Emmanuel Macron pushed ahead anyway, arguing, 'People know that yes, on average, you have to work a little longer, all of them, because otherwise we won't be able to finance our pensions properly.'

Age pensions aside, when you retire is, of course, up to you. Greta Garbo retired at 35; Daniel Day-Lewis at 60; Morgan Freeman has not at 84.

Followers of the so-called FIRE (financial independence, retire early) movement believe that with careful planning and investing, you can retire early in adulthood. 'How to Retire in Your 30s With $1 Million in the Bank' read the headline of an article in *The New York Times* in 2018. To do so, though, you'll need to start saving at a very young age, live somewhere very cheap (such as at home with your parents) and, proponents say, put away 70 per cent of your income, which might be a stretch. (FIRE, unsurprisingly, originated in the United States, most likely with the 1992 book *Your Money or Your Life*.)

'There're too many people providing advice on the ideal time to retire,' says Joanne Earl, a professor of psychology at Macquarie University in Sydney, who has a special interest in retirement planning. 'It's very much an individual choice.' Still, triggers to retire—having enough money, being made redundant or pushing back against a job you no longer like—do not amount to a plan. 'I want people to take a much more considered decision about the exit point,' Earl says. 'A lot of people will leave work and then try to get back in. And sometimes, depending on ageist attitudes and the area that they work in, that may be more difficult than they expect.' In 2020–21, 160,000 people returned to work after previously retiring, according to the Australian Bureau of Statistics. The most-cited reasons for 'un-retiring' were financial for younger people and boredom for older workers.

> The most-cited reasons for 'un-retiring' were financial for younger people and boredom for older workers.

'It's a complex time of life and not an easy change,' observes Julia Barclay, who has been retired for over 20 years. Paul Gottlieb, who stepped away from a career in mining in 2021 when he was 75, is still figuring out what to do with the rest of his life. He and his wife wanted to travel more and spend time with their grandchildren who live overseas but their plans were derailed first by COVID-19 and then by health issues. 'I was never interested in golf and I certainly wasn't interested in men's sheds,' he says.

WHAT IS RETIREMENT, ANYWAY?

Half a century ago, retirement was straightforward: you had ceased paid work, you were relatively elderly (life expectancy was lower) and you were now entitled to a few years of leisure: pottering about the garden, taking an afternoon snooze, maybe, yes, golfing if you were particularly energetic. For a handful of people, there might have been a gold watch.

Yet for most of recorded history, the only people who could afford not to work, at any age, were wealthy landowners—they were pretty much born 'retired'. Everybody else had to toil until they dropped or take their chances on charity, Dickensian workhouses or family

For most of recorded history people had to toil until they dropped.

members. There have been exceptions. During China's Zhou dynasty (1046–256 BC) officials were forced to retire at 70. Later, a word was coined meaning 'retire' (*tuìxiū*), and there is evidence that retirees could expect a few years of leisure, reading books and writing poetry (golf having not yet been invented). The Romans had a kind of proto-superannuation scheme: around 13 BC, Emperor Augustus began granting long-serving legionaries parcels of land in Africa or Gaul (ancient France, home to Asterix), both as a reward for loyalty and to ensure they stayed away from Rome, where they might cause trouble.

The world's first general pension scheme was devised by German Chancellor Otto Von Bismark in the late 1880s (cannily, he put the qualifying age at 70, which few people were expected to reach at the time). In the United States, Union army veterans of the Civil War began to receive pensions from 1890 if they were 65 and not 'unusually vigorous'. Australia followed suit in 1908, when the Commonwealth introduced a national pension of 26 pounds a year or 10 shillings a week (a little more than a third of the wage of a female factory worker) for people 65 and over. The retirement age was later reduced to 60 for women. At the time, being retired was synonymous with being 'old': in 1908, Australian men had a life expectancy from birth of 55, women of 58.

These attitudes persisted. In a 1960s Beatles song, a young man speculates about what life will hold for him and his love interest when he is 64: highlights include weeding a garden and repairing a fuse. The 1990s TV comedy *Seinfeld* was full of retirement gags. In one episode, Jerry Seinfeld jokes that his parents, in their sixties, have moved to Florida, not because they wanted to but because that was the law. Today we're living much longer. In 1965, Australians could expect to reach 71, now it's 83, averaged for men and women. Living to 100 is no longer exceptionally rare. 'Retire' in your sixties and you are not so much crawling to the finish line as embarking on a whole new chapter of your life.

Prudy Gourguechon, a psychiatrist and psychoanalyst in the United States, wants to get rid of the 'R' word altogether. 'As long as we keep using the word retirement or any derivative such as "the new retirement", that whiff of withdrawal, of closure, of endings will linger,' she once wrote. Instead, she proposes the term 'starting older'. It is 'about starting a new and previously poorly defined phase of life, rather than focusing on what you are leaving'.

She adds, 'I just don't think "retirement" is useful, psychologically. It doesn't orient you towards what is really happening in your life, which is that things are changing in very significant ways but you are still, on average, healthy, creative and productive, or can be.'

Certainly, the original definition of retirement—the word is derived from the French *a la retraite*, meaning withdrawn, secluded, out of the way—is well past its use-by date. The current generation of retirees—the Baby Boomers, the youngest of whom turned 60 in 2024—hasn't withdrawn from life. They're busy volunteering, child-minding, consulting, trekking, cruising, driving Ubers, running charities, going to Pink and U2 concerts (Bono, note, is 64) and keen to stay involved. 'Purpose' is their buzzword.

> It is 'about starting a new and previously poorly defined phase of life, rather than focusing on what you are leaving'.

'There are myriad activities to explore, causes to support and new skills to learn, so my "do-list" grows daily,' says Jennifer Ebdon, who loves her seasonal job helping passengers on and off cruise ships at Melbourne's Station Pier. Still, she has pondered the meaning of retirement since leaving work when she was 66.

'The yawning chasm of endless time terrified me and I knew I would need some structure or I would be eating breakfast at dinnertime and living in activewear,' she says. 'There is nowhere to hide from your fears and uncertainties when your brain is no longer engaged with spreadsheets, annual leave and lost socks. What is the purpose of a "full life"? Is pleasing yourself a worthy goal? What is the end game? Why keep gaining skills if you are only going to take them to the grave? What is your value to society now that you are no longer a wage slave?'

Therese Goodacre, a nurse in her early sixties from Orange in New South Wales, is looking towards retirement in the next few years. She has always valued providing an essential service, she tells us. 'For me, retirement means a bunch of empty space and potential loss of work-forged networks. I don't want to fill my retirement hours with meaningless pastimes.' For now, she volunteers at a nursing home to build skills for her life after retirement. 'When I am no longer working, it is important to me that I continue to "contribute". So I am in training to be useful, hopefully.'

IS RETIRING BAD FOR YOUR HEALTH?

We've all heard about some poor sod who went downhill after retiring, and we've concluded that somehow the act of retirement, not just getting older, was to blame. This area has been widely studied, and the evidence is contradictory.

In 2014, the University of South Australia's Dr Tony Daly published a paper that examined the findings of numerous studies into health and ageing. Some studies indicated that retirees were less likely to report good health—both physical and mental. Daly also found several studies that suggested retirement was good for you, including one from 2013 that drew on data from about 7000 Australian households.

As for the belief that retirement could send you to an early grave, Daly noted two studies (one on construction workers and one on people in Norway) that found no connection between retirement and how soon you're likely to kick the bucket. The Norwegian study actually suggested early retirement might be good for men's health in particular. 'Despite contradictory findings, the majority of researchers agree that it is not necessarily retirement, per se, that affects health, wellbeing and mortality,' Daly concluded.

These findings seem to be borne out by a 2016 study into the health and wellbeing of some 25,000 Australians aged over 45. Overall, it found that retirement was linked with positive changes in several areas, including in physical activity, diet, sedentary behaviour, alcohol use and sleep patterns. Lead researcher Melody Ding at the University of Sydney likens the transition to a new year's resolution: 'A critical window where people mentally feel like, "Okay, now I need to do something different about my health because I'm approaching the next chapter in my life".'

Michael Stevenson, who eased out of full-time work two decades ago to focus on amateur motor racing and sailing on Lake Macquarie in New South Wales, agrees. Now nearing 78, he can still do 70 push-ups in a single set. 'The most important thing when you retire is to maintain your fitness,' he says. 'Because if you don't do that, a lot of things you'd like to do in retirement, you can't do.'

> 'The most important thing when you retire is to maintain your fitness. Because if you don't, a lot of things you'd like to do in retirement, you can't do.'

We inevitably lose some mental acuity as we age, particularly in the prefrontal cortex, which controls executive functioning, attention and working memory. Offsetting this increase in 'senior moments' is what's called crystalline, or crystallised, intelligence, accumulated knowledge that can be useful in real-time problem-solving. A review of two major longitudinal studies suggests crystallised intelligence 'shows gains into old age' even if fluid intelligence, which requires fast processing, declines. As in, as we get older, we get wiser. Who knew?

HOW SHOULD YOU PREPARE FOR RETIREMENT?

Yes, a big part of the decision to retire centres around money. Key issues include whether you own your home, how much money you expect to need and whether your savings (such as super), the pension or a mix of the two (subject to earnings and asset tests) will suffice. You need enough, says financial adviser Brenton Tong, managing director of Sydney firm Financial Spectrum, 'and everyone's "enough" is different'.

He advises younger people to start planning immediately: compound interest works miracles if you start saving for your retirement in your twenties, but most of us don't. And he warns: 'For most of your adult life, for most people, money is deposited into your bank account on a periodic basis—month after month. The realisation that [once you retire] no one else will ever put money in your bank account is terrifying!'

Julia Barclay and her husband started seeking out free activities after they retired. 'I also learnt to say no to expensive suggestions,' she tells us. 'It may be necessary to become more frugal. Make a list of the large predictable annual expenses such as house insurance, cars, rates, health insurance. It can be quite a shock to see how much is needed.'

Having enough money is just one of the three main reasons people decide to retire, says Joanne Earl, who has tracked ABS data in this field for a decade. The other two factors usually come unheralded: redundancy or health issues, including caring duties. In other words, plenty of people suddenly end up retired without realising that it has happened. 'I'm saying to people, "Now that you know this, can you be better prepared?",' Earl says.

Michael Stevenson says people who retire need to be prepared both financially and psychologically. 'The second is the harder of the two to come to grips with,' he says. 'One guy I spoke with said he's going to go fishing six days a week. You'd soon get tired of that.'

Terry Kelleher says 'luck and some good management' helped him and his wife, Julie, escape some pitfalls. 'Most people think they can retire in the home they have lived in for the previous ten to 50 years,' he says. 'Also, they assume a retirement plan of some kind can be delayed while they sit back and relax for a few months. That never happens, and months become years, and it's too late. Then either the husband or the wife develops an illness, say cancer.' He was once a driver for community transport in the Southern Highlands of New South Wales. 'The repeated message I heard when picking up clients: "We left it too late to move", and in most cases, they were right.'

Before you retire, think about what your social network will look like when you no longer have work colleagues, says Julia Barclay. 'If possible, plan ahead, well ahead, a year or more, to develop new friendships, new hobbies,' she advises. 'Suddenly, the weekly contact with large numbers of people will disappear.'

Judith Venables, however, doesn't think you need a specific plan. 'My view is that planning effectively relies on knowing what will happen in the future, which is something no one knows,' she says. 'We gather information until we're ready to make our gut decision based on the information we have chosen to prioritise. So if we feel like doing it and it has a reasonable chance of success—go for it and deal with whatever comes next.'

HOW DO YOU COUNTER 'RELEVANCE DEPRIVATION SYNDROME'?

In Japan, some retired workaholic men are known as *nureo-chibazoku*, or 'wet fallen leaf', reports *The Economist*. Friendless and devoid of hobbies, they 'follow their wives around like a wet leaf stuck to a shoe'. Australians might experience it, less poetically, as a feeling of worthlessness that descends after the initial thrill of leaving work wears off. Who am I? What's my purpose now?

Or not. 'People said I would be bored, I would have relevance deprivation, but none of that has come to pass,' says Charmaine Moldrich, who retired as chief executive of the Outdoor Media Association in Sydney in 2023. She's still enjoying the first flush of retirement: working in a communal garden she started in her back lane, exercising, eating better, getting healthier, travelling, cooking, cleaning, hanging out with friends, babysitting, being more community spirited—'doing all the things I did in a hurry a bit slower and with joy'.

Others transition into retirement by working part time, volunteering or building a 'portfolio career' with several small income streams but no single big commitment.

The key is redefining yourself, says Theo Van der Veen, who retired from a senior education role in Newcastle at 58 and loves mountain-bike riding. He was offered consulting work but realised a seismic change had already occurred in his life. 'Your use-by date expires virtually on the day that you retire,' he says. 'My advice was about as welcome as ex-prime ministers weighing in on current political issues.' Today, he tells us, 'I am no longer seen as Theo the school director but rather Theo the bike rider, photographer, cabinet-maker, father, grandfather.'

Bob Doyle, who retired at 60, took a less demanding job driving a minibus for children with a disability, leaving him with plenty of energy to do other things in North Avoca on the NSW Central Coast. He took up cricket again when he was 70. 'The one piece of advice I have not forgotten is that for the first 12 months or so, you will be on a high, but expect to feel deflated after that if you haven't got some firm routines in place based on getting to finally do all those things you had to put on hold when working,' he tells us. 'More than 10 years after "retirement", I've had two other jobs and started my own business. I'm now entering a different phase not involving work, but loving the possibilities.'

'For the first 12 months or so, you will be on a high, but expect to feel deflated after that if you haven't got some firm routines in place.'

But a 'pre-tirement' gig—where you wind back work but don't ditch it altogether—can be a mixed bag, warns Julia Barclay. 'Useful but also frustrating. Professionally you are neither in an active position, regarded as an equal, nor available to be more spontaneous and enjoy the freedom.' She suggests, if possible, organising part-time work into blocks of full-time work and blocks of 'not available'.

Indeed, after Judith Venables left her career as a social worker, she found work at another organisation. Although they were welcoming and the work was interesting, the experience cemented what she already knew: 'All the bureaucracy, the this and the that, I didn't really need this anymore.' Now, she is a volunteer wildlife carer for birds, nursing the occasional tawny frogmouth back to health.

'Don't just sit on your bum and wait for things to happen,' says Lawrie Mann, who enjoys sailing weekly. 'Go out and make something happen.'

Alison Muirhead, who retired from teaching at 56, has had a positive experience as a volunteer tour guide for Brisbane City Council alongside her husband of 50-plus years, Ian. 'First up, pick a partner with the same interests as you,' she advises. 'In our case, it has included Brisbane's history, memoir and fiction writing, environmental volunteering and bird watching.' Note our retired arts administrator has found there's hot demand for volunteer roles. His advice is to set something up with your work contacts—before you retire. 'We've got a lot of competition out there volunteering, with all of us Boomers out there and more coming online that are all wanting to do stuff.'

Jeff Broderick, from Croydon in Melbourne, who retired from his job in the superannuation industry 10 years ago, has a simple rule for retirement: have something to do. 'Wake up in the morning and not wonder, *What will I do today?* but *What on my long list will I do today?*' he says. 'Be active. Golf, tennis, run, walk, Pilates, whatever. Do it with other people. Make new friends. Hey, make new enemies. It gets the blood flowing.'

24

WHAT MAKES CAR CRASHES SO DEADLY?

Vehicle collisions kill an average of three people a day in Australia. Bad driving is often blamed—but other factors can play a part.

Patrick Hatch

Jessica Zaghet had just finished her first semester as a primary school teacher in eastern Victoria when a car crash changed her life. 'I loved the kids and I loved being a teacher but it was full-on. So I was feeling relieved, thinking, *Finally, two weeks off,*' she recalls. She was 24 and leading a busy life, playing netball and socialising with a tight group of friends. She said goodbye to her students in the town of Churchill and started the 20-minute drive on a back road to her home in Traralgon.

What happened next isn't completely clear. Zaghet thinks her exhaustion may have led to a moment of inattention. The last thing she remembers was seeing the truck in front of her and trying to swerve back into her own lane. The collision crushed Jessica's small car. The truck driver was not injured but Zaghet was airlifted to a hospital in Melbourne in a critical condition. 'Everything changed in one moment,' she says.

On average, road crashes kill more than three people and send another 100 to hospital every day in Australia. Twenty-five people a day sustain life-threatening injuries. Crashes are a leading cause of death among children aged between one and fourteen, and the second-biggest killer of people 15–24 years old. Australia's road fatality rate (4.54 deaths per 100,000 people) in 2022 was more than double the rates in the world's safest countries, Norway (2.1 deaths per 100,000 people) and Sweden (2.2), placing us 18th of the 31 nations in the Organisation for Economic Cooperation and Development (OECD) that were studied. Globally, road collisions kill 1.3 million people each year.

But while drivers are often the focus of blame, other factors contribute to accidents, such as the cars we drive and

Australia's road fatality rate in 2022 was more than double the rates in the world's safest countries.

the roads we hurtle along. What part do they play? Why do people die on our roads? And what can be done about it?

WHAT MAKES CAR CRASHES SO CATASTROPHIC?

Zaghet didn't regain consciousness for three months. She was in a rehabilitation centre and didn't know where or who she was. Gradually, she regained her memory and her ability to speak and started the painful process of learning to walk again. Eight months after the crash, she was finally discharged.

'We're actually quite fragile creatures but, in our minds, we don't see ourselves that way,' says Dr John Crozier, a Sydney trauma surgeon who has treated thousands of car-crash casualties over his three-decade career. 'The forces imposed on the body in crashes, even at low speed . . . are well past the threshold where organs will be torn apart or shattered, and limbs will be torn apart and shattered.'

Zaghet knows that some people will have little sympathy for her. After all, she had drifted to the wrong side of the road. She decided to share her story because she wants others to realise how easily the same thing could happen to them—even people who are confident they are 'good drivers'. 'I always did the responsible thing,' she says. 'I was always careful and still this happened to me. It could happen to anyone.'

Imagine driving down a country road at 80 km/h. Your car is full of kinetic energy—the energy of motion—created by your speed and the mass of your vehicle. If you hit something, that energy has to go somewhere: it can generate heat and noise, be absorbed by the car frame, which will crumple and fold, and be transferred to the occupants of the car or anyone you hit.

There's a split second after impact when the car will have stopped but the driver's body will keep moving.

Bodies are loaded with kinetic energy too. There's a split second after impact when the car will have stopped but the driver's body will keep moving until it is halted—by a seatbelt, an airbag, a steering wheel, a windshield or another part of the vehicle.

There's a third impact, too: a body stops moving but its insides don't. Vital organs can hit bones with enough force to rupture. The brain can sustain fatal injuries from the force of hitting the inside of the skull. 'So you do see patients slumped forward dead, with no obvious signs of injury to the face or other scalp structures externally, but that de-accelerating force was such that the brain has been extensively damaged,' Crozier says.

Brain injuries are the most common cause of death in a crash, followed by aortic rupture, in which the body's main artery is dislodged from the heart. Zaghet's injuries were typical of ones sustained in a head-on crash at speeds above 80 km/h: a severe traumatic brain injury and haemorrhage, two collapsed lungs, fractures to her legs, ribs and spine, and damage to her stomach muscles and organs. Ruptured guts, bowels and bladders and paralysis are common, too. Such injuries, says Crozier, can be 'the price people pay to survive a crash.'

WHAT PART DO DRIVERS PLAY IN CRASHES?

In almost every car crash, somebody has done something wrong: be it a dangerous or illegal activity or a simple misjudgement of distance or speed. Queensland Police says most road deaths are caused by the 'fatal five': speeding, distraction, driving under the influence of alcohol or drugs, fatigue and failing to wear a seatbelt.

Safe drivers aren't necessarily ones who are most skilled, says Narelle Haworth, a road safety researcher at the Queensland University of Technology in Brisbane. 'Often confidence and skill actually contribute to risk taking and to not only illegal but unsafe behaviours,' she says. 'What we need is alert and compliant drivers.' Of the people killed in crashes in 2023, the biggest group was aged between 40 and 64 (387 people) followed by those aged 26–39 (281) and 17–25-year-olds (244).

> Safe drivers aren't necessarily ones who are most skilled ... Often confidence and skill contribute to risk taking and unsafe behaviours.

Drink-driving laws, better car design and lower speed limits have helped cut road deaths by a third since 1970. Victoria led the way in the world when it made wearing seatbelts mandatory in 1970, and other states quickly followed. Still, 1266 people died in road crashes in 2023, including 158 pedestrians, 254 motorcyclists and 36 cyclists—a 7.3 per cent increase from 2022. Progress in reducing road deaths hasn't just stalled; since the COVID-19 pandemic, road deaths have increased.

Almost one in five drivers killed in crashes was over the legal blood-alcohol limit of 0.05 (that's 0.05 grams of alcohol to every 100 millilitres of blood). Drug use is estimated to contribute to as many as a third of fatalities. In 2021, 21 per cent of vehicle occupants killed in crashes weren't wearing a seatbelt, according to federal government data. Meanwhile, the NSW Transport Department estimates that speeding is a factor in about 40 per cent of road fatalities and a quarter of serious injuries. Driving 5 km/h over city speed limits

Driving 5 km/h over city speed limits and 10 km/h over country limits roughly doubles the risk of a crash.

and 10 km/h over country limits roughly doubles the risk of a crash, according to Queensland's Centre for Accident Research and Road Safety.

The array of driving distractions has only increased over time. Haworth and other researchers put cameras in 346 cars in Victoria and New South Wales in 2015 and recorded their owners making a combined 194,961 trips. Drivers spent about 45 per cent of their time doing 'secondary tasks': adjusting seatbelts, fiddling with radios or air-conditioning, eating and drinking, touching their phones (either in their hand or in a holder), reaching for objects and doing 'personal hygiene tasks'. Analysis of 43 hours of the recordings found 95 dangerous incidents, including eight drivers swerving out of their lane, five failing to indicate and one failing to give way to a pedestrian.

'People drive as they live,' Haworth says. 'A lot of the pressures of everyday life actually affect your driving. If you're worried . . . or you're feeling lonely, then you're more likely to use your phone more often and then you're more likely to be on the phone while you're driving.'

Sometimes driver behaviour is so bad it becomes criminal: for example, in the year to July 2023 in Victoria, 36 people were found guilty of dangerous driving causing death and 13 more were sentenced with the more serious charge of culpable driving causing death, according to the state's Sentencing Advisory Council. In New South Wales, 2270 motorists were found guilty of dangerous or negligent driving, which cover a broader range of offences, over the same period. In both states, 80 per cent of drivers charged with those offences were men.

WHY ARE SOME ROADS MORE DEADLY?

'We can't make the driver perfect,' says Michael Fitzharris, associate professor at the Monash University Accident Research Centre in Melbourne. Police enforcement makes a significant difference to behaviour, he says, but stopping people from ever doing the wrong thing is impossible— which is why looking at roads and cars is also important.

Drivers, cars and roads where crashes happened were all put under the microscope by Fitzharris and colleagues in a 2020 study of 393 crashes, involving 400 drivers hospitalised for injuries. Forty-five per cent of the drivers met the 'safe' criteria: they weren't speeding, using a phone, under the influence or excessively drowsy. But just 10 per cent of the vehicles passed the safety benchmark, which required them to have front and side airbags and electronic stability control, and to have received a five-star safety rating from the Australasian New Car Assessment Program (ANCAP).

Even the safest cars can protect passengers from injuries in a front-on crash only at speeds of up to 70 km/h, and in side-on crashes at speeds of up to 50 km/h. So the study gave a safety tick only to roads that had speed limits below those thresholds or had infrastructure such as roadside barriers, separated lanes, turning lanes and roundabouts to prevent those kind of crashes. Only a quarter of the roads they examined met these criteria.

'People traditionally have reverted to blaming the driver, but we see clearly in our research that even if compliant drivers are doing everything right, the vehicle and the broader road infrastructure are just not supporting them to be safe,' says Fitzharris. Only eight crashes in the study involved safe drivers, cars and roads—and none of them led to serious injuries.

Chances of surviving being hit by a car.

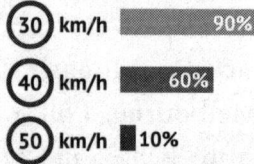

30 km/h 90%

40 km/h 60%

50 km/h 10%

*Based on young adult pedestrians

Vehicle 1*   Vehicle 2

Chances of surviving a side-on crash.

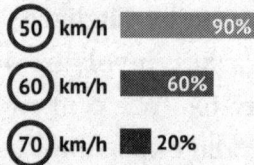

50 km/h 90%

60 km/h 60%

70 km/h 20%

*Based on Vehicle 1 speed

Vehicle 1*   Vehicle 2*

Chances of surviving a head-on crash.

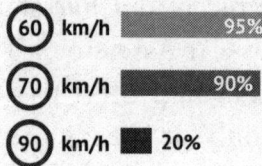

60 km/h 95%

70 km/h 90%

90 km/h 20%

*Both are light vehicles of similar size and mass, travelling at the same speed

Jamie Brown, based on data from Per Wramborg, Swedish National Road and Transport Research Institute, 2005

The stretch of country road where Zaghet crashed has all the qualities that another road-safety expert, Rob McInerney, is on a mission to eradicate: it is narrow and tree-lined, has a 100 km/h speed limit, in places has only a white line down the middle and doesn't have a rumble strip to alert drivers when they veer off course. While most people would drive on the road at 100 km/h without a second thought, McInerney suggests a thought experiment: close your eyes and imagine sitting in a deck chair on the side of that road while a truck zooms within centimetres of you at 100 km/h. 'How safe would you feel?' he asks.

About two-thirds of Australia's road deaths happen in regional and remote areas where roads like this are common. 'The road with killer roadsides or the road with no sidewalks or safe crossings for pedestrians—every one of those has the risk of death built-in,' says McInerney. 'So we shouldn't be surprised when a crash happens.'

McInerney heads the International Road Assessment Programme (iRap), a charity that works with bodies such as the United Nations to eliminate dangerous roads around the world. 'We're all humans, we all make mistakes, but we shouldn't pay with our life for that mistake,' he says. 'Our buildings aren't built without balcony rails, we have a door on the elevator shaft, but we can have a 10-metre drop on a roadside and think it's normal.'

iRap has rated the safety—up to five stars—of more than a million kilometres of roads in more than 100 countries. It looks at design features such as footpaths and streetlights for pedestrians, bike lanes for cyclists and the width of the centre line that divides motorists driving towards each other. With each additional star, the risk of serious crashes drops by half.

Based on iRap's most recent look at 100,000 kilometres of Australian roads, we spend 58 per cent of our driving time on roads rated three stars or better. Updated ratings, due in 2025, are expected to show a significant improvement. The federal government's National Road Safety Strategy wants 80 per cent of travel on roads that have a speed limit of at least 80 km/h happening on roads with at least a three-star rating by 2030. In countries including Denmark, Ireland, the Netherlands, Spain, Britain and Switzerland, more than 85 per cent of travel is on roads that are at least three-star.

The most common type of deadly crash in Australia involves a single vehicle running off the road and crashing into a tree or pole or rolling down an embankment (making up 37 per cent of fatalities in 2021) followed by crashes at intersections and head-on collisions (15 per cent each). McInerney says this shows that more than half of Australia's road deaths could be addressed with infrastructure improvements: clearing roadsides, building barriers, separating two-way traffic and adding merging lanes or roundabouts, which force drivers to slow down.

The most common type of deadly crash in Australia involves a single vehicle running off the road.

WHAT SAFETY FEATURES SHOULD NEW CARS HAVE?

Until recently, cars have been built to protect occupants from the effects of collisions with airbags, seatbelts and 'deformation zones' that crumple in a way that absorbs kinetic energy. But new technologies are helping cars prevent crashes too (we're not talking about self-driving cars—it's not clear if or when they will become a common sight on our roads).

Autonomous emergency braking uses radars to detect an impending collision and slam on the brakes. Other so-called advanced driver assistance systems include blindspot detectors, automatic lane centring and adaptive cruise control. All new vehicles in Australia will be required to have car-to-car autonomous emergency braking from March 2025 and car-to-pedestrian braking from 2026. Mandatory safety features are set out in the federal government's Australian Design Rules for new vehicles, which cover everything down to windscreen wipers. Most new cars are also tested by

ANCAP; its star rating system, which isn't mandatory, aims to educate consumers and encourage carmakers to exceed minimum safety standards.

While smart technology promises to prevent some crashes, safety experts are alarmed by the growing size, weight and shape of vehicles. A study released in 2023 by the US Insurance Institute for Highway Safety found that large North American pick-up trucks, SUVs and vans with front ends one metre high or taller—increasingly common on Australian streets—are 45 per cent more likely to kill a pedestrian than vehicles with shorter front ends. Even 'small cars' are bigger than they were a few decades ago— and all vehicles will get considerably heavier as they go electric.

WHAT CAN BE DONE TO REDUCE CAR CRASHES?

When Victoria experimented with increasing speed limits on rural roads and freeways from 100 km/h to 110 km/h in 1987, crashes that resulted in fatalities and serious injuries jumped 21 per cent. They fell 19 per cent when the state restored the old speed limits two years later. Deaths also fell about 20 per cent when state governments reduced the default speed limits for urban areas from 60 km/h to 50 km/h around the turn of the century.

Road safety experts say Australia needs to cut its speed limits further. 'That has two effects: one, it gives the drivers more ability to make safe decisions because things are just that little bit slower,' Fitzharris says. 'But also, when a crash occurs, the vehicle has much more chance to protect them from serious injury and death.' McInerney agrees that speed limits on quieter country roads, where infrastructure upgrades might not be justified, should be reduced from

100 km/h to 80 km/h. Default rural limits are 80 km/h or 90 km/h in much of Europe, and places such as Britain are rolling out 30 km/h zones on many streets in local communities.

Why do speed limits have such a significant effect on road safety? Consider the formula for kinetic energy: $KE = \frac{1}{2}mv^2$. That is, a moving object will carry kinetic energy equal to one half of the product of multiplying its mass (m) by the square of its speed (velocity, or v). A seemingly small increase or decrease in speed produces a big change in the potentially lethal energy. For example, when a two-tonne car slows from 85 km/h to 75 km/h, the speed has decreased 12 per cent. But the kinetic energy falls 20 per cent. Studies show this 10 km/h reduction will reduce the chance of a serious collision by 29 per cent.

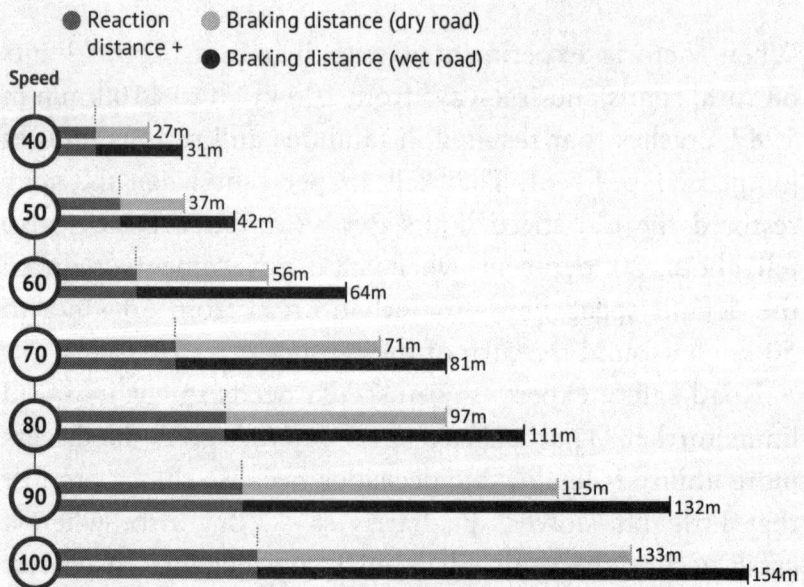

While you react, your car travels many metres, and it keeps going even after you hit the brakes.

Jamie Brown, based on data from Transport for NSW

'Speed is the big-ticket item that I think we really need to look at still,' says Professor Rebecca Ivers, who heads UNSW's School of Population Health in Sydney and has worked extensively on road safety. 'But it's a cultural thing—we all think we've got a God-given right to get to where we want to go as quickly as possible.'

'It's a cultural thing—we all think we've got a God-given right to get to where we want to go as quickly as possible.'

Ivers takes the discussion of road safety to a more fundamental level: does everyone need to be driving in the first place? 'You are always going to have recidivist drink drivers,' she says. 'So can we make sure that someone who's got an addiction problem doesn't feel like driving is the only option available to them because there is a bus or some other form of public transport? We're not going to attack the road injury issue unless we get people out of cars and into other forms of transport.'

State and federal governments have all committed to cutting road fatalities by half and reducing serious injuries by 30 per cent by 2030. The National Road Safety Strategy says it will invest in road upgrades, encourage safer vehicles and collect better data about road trauma. In 2023 and 2024, the strategy focused on trucks, which are involved in 17 per cent of all road deaths.

Twelve years after her crash, Jessica Zaghet lives independently but chronic pain and other physical and psychological injuries mean she is unlikely ever to work again or to live the active life she once had. Instead, she finds moments of happiness watching movies at home with friends and visiting old school friends and their children. As a volunteer with the road safety education charity Amber Community, she talks to people about the effects of road trauma. 'I'm just living whenever I can, with the life that I have left,' she says.

If there's one thing that she wants people to know before they get behind a steering wheel, it's this: 'You have to be extremely careful and cautious when driving at all times. I just want people to know about the damage they can do when they get in a car.'

25

WHAT IS A GIFTED CHILD?

They're sharp and curious but 'gifted' kids can have a tough time. What's life like when you've won the intellectual lottery?

Angus Holland

Isabella, a slip of a thing with bright, curious eyes, is like any other 11-year-old in many ways. She lives with her mum and dad and big sister in a suburban house filled with games and puzzles, plays sport, goes to the movies with her friends and loves her dog, an adorable barky spaniel-something-cross resembling a mop that's lost its pole.

In conversation, though, it quickly becomes apparent that Isabella—not her real name, for reasons that will become clear—is a bit different to other kids her age, thanks to an adult-like vocabulary and an IQ that puts her into the top 1 per cent of her age group. 'I am highly gifted,' she says. 'But I don't think of it at all as a gift because the struggles of it are quite hard most of the time.'

Indeed, being 'gifted' in Australia is a mixed blessing, as Isabella's parents have learned. Educators don't always know what to do with a child who is academically years ahead of their contemporaries. Other parents might dismiss a gifted child as the product of middle-class hot-housing, manufactured to somehow gain an unfair advantage. There's not even a common understanding of what 'gifted' means. It is often lumped in with 'talented', which, while related, is entirely different. Even the word itself can be seen as problematic, perhaps implying that some children are more special, lucky or 'blessed' than others—anathema in our supposedly egalitarian society.

So, what is giftedness? And what is life like for children such as Isabella?

WHAT DOES 'GIFTED' MEAN?

In popular culture, the gifted child is often portrayed as a curiosity: bookish, nerdy and capable of impressive intellectual stunts such as reciting pi to a hundred decimal places. Prodigies do pop up in real life. Mozart wrote *Trio in G major* when he was five. US philosopher Saul Kripke had taught himself ancient Hebrew by six. Mathematician Ruth Lawrence went to Oxford University at 12 and became its youngest graduate at 13. Singaporean wunderkind Ainan Celeste Cawley passed a Year 10 chemistry exam when he was seven and, yes, could recite pi to 521 decimal places when he was nine. Which is a lot.

The notion of 'giftedness' in children, at least as we understand it today, dates back to 1905, when Frenchmen Alfred Binet and Théodore Simon devised the first modern IQ test to reveal a child's intelligence compared to others of the same age. Controversial US psychologist Lewis Terman built on Binet's work with a long-term study of children with high IQs—his doctoral dissertation was titled, somewhat insensitively, 'Genius and stupidity: A study of some of the intellectual processes of seven "bright" and seven "stupid" boys'.

Yet, a century later, what constitutes giftedness is still being debated. Some educators deny its existence entirely or groan that it is no more than a fantasy of pushy parents. Some take the charitable view that all children have a particular gift of their own. Others conflate giftedness with talent when the two are often discrete: a gifted child may have been born with great potential but not have explored or displayed it. (Canadian psychologist Francoys Gagne distinguished giftedness from talent, offering explanations of how natural abilities can be developed into specific skills.)

Another common definition of giftedness is simply as demographic rank—the top X per cent of children, measured by academic achievement, in a given cohort. In Australia, that figure is typically 10 per cent, which would mean two or three children in any particular classroom are 'gifted'. More exacting, Singapore originally steered just a quarter of a per cent of its brightest children into its Gifted Education Programme, which was modelled on an Israeli system from the 1980s. The Singapore school system has since expanded the entry criteria to the top 1 per cent of students. In the United States, a gifted child is commonly defined as one who performs at, or shows the potential to perform at, a remarkably high level in an intellectual, creative or artistic area, compared to children of the same age.

Yet psychologists are trained to limit the 'gifted' tag to the top 2 per cent as measured by currently recognised IQ tests, says veteran Melbourne child psychologist Judy Parker. (It should be noted that IQ tests are a somewhat controversial and imprecise tool for measuring intelligence but are probably the best we have for now.)

'There are no agreed definitions of giftedness and talent,' admitted the report of a Victorian parliamentary inquiry in 2012. It settled on the catch-all 'young people with natural ability or potential in an area of human endeavour'. Nor can 'giftedness' necessarily be quantified, the inquiry noted.

'A gifted student may have exceptional abilities in some areas but be average, or even below average, in others.'

'It is impossible to paint a single picture of a gifted student ... Gifted students are not a homogenous group. They come from all socioeconomic and cultural backgrounds. Their gifts may be across a vast array of different domains, from academic to creative to interpersonal. A gifted student may have exceptional abilities in some areas but be average, or even below average, in others.'

HOW DO YOU RECOGNISE A GIFTED CHILD?

It's not always easy. As many teachers will tell you, the number of parents who declare their child gifted vastly outnumbers the reality. Some truly gifted children mask their intelligence to fit in with their peers while others initially present with learning behaviours that can suggest autism or attention deficit hyperactivity disorder (ADHD) and turn out to have an IQ in the gifted range. Many of the parents we spoke with had their children assessed by psychologists after noticing unusual or odd behaviours, when they struggled to fit in at school or when they became disruptive in the classroom.

A handful of children—those with an IQ found in only one in 10,000 people—likely think in ways the rest of us cannot begin to comprehend. They may be top of their class or, as physicist Stephen Hawking was as a child, seem to be away with the fairies.

For argument's sake, let's describe 'gifted' as possessing an intellect that processes information faster, learns concepts quicker and retains knowledge more readily than most, allowing it to explore ever more complex ideas and to make increasingly insightful connections. 'They learn rapidly, they have an excellent memory and they reason well. These are the sorts of characteristics I'd note,' says Judy Parker, who has spent much of her career testing for giftedness.

For some of these children, maths or language skills come early and easily; reading might come naturally, long before tuition; a child hungry for knowledge might temporarily nurture an alarmingly deep interest in an esoteric subject such as black holes or London bus routes; typically, they will have a precocious vocabulary full of spelling-bee words.

IQ tests that psychologists use to identify giftedness include the Wechsler Intelligence Scale for Children, which measures verbal comprehension, visual-spatial abilities, fluid reasoning, working memory and processing speed. IQs (again, somewhat controversially) are plotted on a bell curve, with most of us in the middle, clustered around an average of 100 and exponentially fewer of us at the edges. An IQ of 160 on this scale is rare, translating to one person in 10,000 or more. An IQ score is not an absolute number, though. A child might test differently in different circumstances, such as if they are tired or hungry.

Some of the parents we spoke with are happy to share their child's IQ score; others prefer to speak in terms of 'moderately' or 'profoundly' gifted, or where their child sits in a percentile of the population.

Helen, a mother we spoke with, laughs as she remembers her young son's precocious behaviour. He was able to recite the alphabet as a two-year-old, a recognised signifier of exceptional giftedness, according to US educational author Deborah Ruf. Then a family friend told him, jokingly, that he wouldn't truly 'know' his alphabet until he knew it backwards. 'And so he just rattled it off backwards without giving it a second thought,' says Helen. Several years later—when her son was 11—Helen took him to be assessed after he had trouble fitting in at school. His IQ score was improbably high.

Isabella wasn't doing algebra in her cot but she did have proper conversations with her mum at 18 months. Her family later had her psychologically assessed, thinking she might have an autism disorder. It turned out she had an extraordinarily high IQ. A decade on, she has just aced a test on a book that she'd read only the start of because she had 'managed to infer the rest'. The *Barbie* movie was funny, she tells us, but none of her friends understood the adult-oriented jokes, 'which was really awkward'. And even though she only recently started learning Mandarin, she no longer needs to read the subtitles on Chinese films because 'I can kind of understand it now', she says, proudly reciting a long phrase.

Yet a perpetually active brain like hers has its drawbacks. 'Birthdays are not easy for Isabella,' confides her mum. 'By the time we come to the birthday, she's already calculated how many days she has left to live, how many days the dog has left to live, grandma has left to live. She's able to reduce the bigger picture into the main facts that aren't always wonderfully positive.'

Siblings will typically have similar IQs, though their 'gifts' might be quite different. And high intelligence is not always demonstrated at a young age: Hawking was reportedly a late developer, never reaching the top half of his class. Albert Einstein didn't speak in full sentences, according to some of his biographers, until he was five. (Perhaps he had better things to think about.) Robert Oppenheimer, who led the World War II program to build an atomic bomb, was probably a gifted child: he described himself as 'an unctuous, repulsively good little boy', skipped ahead in school, read widely and later taught himself Sanskrit. He famously quoted from the ancient Hindu text *Bhagavad Gita* after the first successful nuclear weapon test in 1945: 'Now I am become Death, the destroyer of worlds.'

CAN YOU 'MANUFACTURE' A GIFTED CHILD?

Some argue that with the right educational techniques, many, if not most, children can reach the level of a mildly gifted child. The idea is either that all children are potentially gifted, or that gifted children are the product of coaching and parental effort, not some kind of magic they carry from the womb.

'Research is clear that brains are malleable, new neural pathways can be forged and IQ isn't fixed,' writes Wendy Berliner, co-author of *Great Minds and How to Grow Them* from 2017. 'Most Nobel laureates were unexceptional in childhood.'

> **'Research is clear that brains are malleable, new neural pathways can be forged and IQ isn't fixed.'**

Much more important, this argument goes, are perseverance, effort and quality teaching.

Then there's the belief that all children are gifted in their own way. 'The great teachers and the great schools find the gifts in every student,' agrees Deborah Harman, who has worked as an educator for 45 years and is one of the leaders in Victoria's Accelerated Learning Program. 'All students—especially those who are gifted—need to feel a sense of belonging to their classes and their peers.'

And yet, some children—the alphabet boy, for one—seem to be different right out of the blocks, behaving in ways no parent could have confected.

Elissa McKay, the mum of a gifted boy, Finn, grew so frustrated with the myths surrounding giftedness that she compiled her own 'primer' that was widely shared in internet forums frequented by parents of gifted children (largely to share tips, schools advice and war stories, and to help answer the perennial question: is my child gifted?).

We met Elissa at her home on the leafy outskirts of Melbourne, where Finn has just celebrated his birthday, which meant the long-awaited acquisition of a new yo-yo (yes, the craze has come around once again) and the surprise adoption of a kitten.

Like many gifted children, Finn was reading fluently three years before he started school; when he did start, he was reading at high-school level, says Elissa. But he had something of a mixed educational journey until he skipped ahead two years at school (three years for maths). Before then, she says, 'his reports were really mediocre'.

It's a myth that gifted children need no educational support, she tells us. It's a subject she addresses in her primer, along with the notions that gifted children are best treated just like their peers (she says they shouldn't be) and that they eventually 'level out' (she says they never do).

Julia Lewthwaite of Sydney recalls taking her son Andrew Attard for a check-up at a child and maternal health centre when he was six months old. 'He started pulling puzzles from a shelf and actually completing them. The nurse basically said to me, "I think your child might be gifted".' By age three he was reading, 'and that's where the real craziness started', she says. 'He wanted to go walking down all the streets, reading all the street signs, reading all the house names, the house numbers.' Tests revealed that Andrew was in the 'profoundly gifted' range, about one in 30,000 students. At 15, he became the youngest student to complete the NSW Higher School Certificate in 2023.

If you want to 'manufacture' a gifted child, you should probably start with two smart parents. Many of the families we spoke with professed to be surprised when their offspring tested in the gifted range, but it usually turned out they had a rocket scientist for an aunt or an uncle.

'Both my parents, I'm sure, are profoundly gifted,' says another mother of gifted children. 'My parents met at Oxford (university), my husband and I met at Oxford, there isn't anyone without a PhD, you know. It's just snowballed in our family.' Helen, meanwhile, suspects that her husband, who went to a selective school, is gifted and that she might have been herself, although her focus as a child brought up by a struggling single mother was survival, not excelling academically. 'I had a very different journey in that respect,' she says.

WHY IS THE IDEA OF GIFTEDNESS SOMETIMES CONTROVERSIAL?

'Neither teachers, the parents of other children, nor the child peers will tolerate a wunderkind.'

As US cultural anthropologist Margaret Mead once observed: 'Neither teachers, the parents of other children, nor the child peers will tolerate a wunderkind.' Every parent we spoke with was extremely wary of being portrayed as pushy, boastful or deluded. Some preferred to reveal just their first names; others asked to remain anonymous (although Isabella is the only pseudonym we have used).

'People talk,' says Kate, whose daughter Libby, 10, ploughed through all of her entry-level books almost immediately when she started school and had to go up a year. 'A mum might come up to me and say, "Oh, my daughter said how fantastic Libby is at maths" but the natural inclination is to downplay it.' All kids just want to fit in, she observes. 'I think a lot of people make the assumption that kids that are really smart must be egotistical and they must walk around thinking they're smarter than everyone else but it's usually the opposite. They usually try to hide their true self.'

And while some parents, unsurprisingly, can't help but show a little pride in their gifted children, others resolve to treat their offspring as 'normal'. One mum whose children have tested with IQs in the 99.9 percentile is more interested in raising them to be good citizens than encouraging what she calls 'party tricks'. 'They didn't teach themselves to read or speak Russian or anything like that,' she says. 'They're just normal kids who pick things up quickly.'

Giftedness can provoke discomfiting notions of a handful of children belonging to an elite group that, by definition, excludes the 'non-gifted' masses. Indeed, early IQ tests were developed with more than a dash of eugenics—the debunked pseudo-science that flourished in the Victorian and Edwardian eras. One early test had the stated goal of 'curtailing the reproduction of feeble-mindedness and the elimination of an enormous amount of crime, pauperism and industrial inefficiency'.

US psychologist Lewis Terman, who drew on the work of Alfred Binet to produce the first versions of today's Stanford-Binet test, was a fan of eugenics. Yet he was also determined to challenge the contemporary prejudices against gifted children, particularly that they were physically weak and anti-social. In an extraordinary long-term study, he recruited 643 children of various ages and published five volumes of findings over 35 years. He concluded that the gifted children, whom he called 'Termites', thrived in both their professional and personal lives, by and large.

Concerns about how IQs and educational potential are tested continue to inform gifted education in the United States. New York City scrapped a standardised test that, according to The New York Times, 'foreclosed opportunity for thousands of Black and Latino children' because of its biased cultural references.

'I know IQ tests get a really bad rep,' says Elissa McKay. 'But they're highly accurate at measuring what they measure. They don't measure potential for success. They don't measure for potential for happiness. But they do measure a small subset of characteristics associated with intellectual potential.'

Psychologist Judy Parker also believes they are an effective tool. 'It's the best quick and comprehensive and individualised assessment, if done by a psychologist experienced in the field,' she says.

WHY DO SOME GIFTED CHILDREN UNDERACHIEVE?

If your child is so smart, how come they're not top of the class? This was a familiar refrain to many of the families we spoke with. Many have spent years persuading educators to let their child skip a grade or to be given more advanced work, only to be told their supposedly gifted child wasn't even keeping up with their current grade levels.

Gifted children start school with enthusiasm but quickly tire of it because they are forced to crawl ... when they should be running.

These children are not completely otherworldy: some stuff still has to be learned. Hawking, according to one biographer, did so little revision at Oxford that he decided to focus on theoretical questions in his final exams rather than get caught out on the ones that required recalling facts. One explanation for their poor performance is that gifted children start school with great enthusiasm but quickly tire of it because they are forced to crawl through the curriculum when they should be running. 'Just because you are highly able does not necessarily mean that you will achieve to a corresponding level,' says Jae Jung, an associate professor at UNSW Sydney who has a research interest in gifted children. 'One of the reasons we have so much of this underachievement happening is that teachers are not trained in giftedness, which means that these gifted kids are not being looked after. They're being given content that's, say, three years below their level of capability so they're bored, they're not going to do the work, hence they're not achieving to their full potential.'

According to Parker, 'virtually no' training for teachers in giftedness is offered at the undergraduate level. She also says a number of studies have shown that teachers are not very capable at detecting the gifted children in their classrooms.

Isabella tells us at her former school she would finish an hour's work in six minutes, then stare at the ceiling for the rest of the class. Another mother worries that her gifted daughter—whose interests range from teleportation to the origins of the Christian calendar—is losing her love of learning. 'They disengage and tune out and don't find it interesting,' she says. 'We want her to strive for more and not think, *Oh, this isn't for me, I'm not that smart because I'm not doing well at school*, and just give up.'

'It's still about engagement,' says Harman, who is principal at Melbourne's Balwyn High School. 'It's still about inspiring them.' Often, she says, poor performance is tied to a student's low self-esteem or their relationships with classmates. 'There can be kids who underperform because they don't want to reveal how talented they are.'

'Twice exceptional' children might show signs of giftedness but also have, say, ADHD or autism spectrum disorder, which can make it difficult to learn at school. A study of US members of Mensa, the international society whose price of admission is a high IQ, found that high intelligence often co-exists with super-sensitivities such as strong emotions and reactions to stimuli.

'What's extremely difficult about having those two things combined,' says a mother with a gifted daughter, 'is that children like this, their natural intelligence kind of floats them through at a decent level. Not at a high level, but at a level that is still above a lot of other kids, which is what we're experiencing. And the school will say, "Well, they're doing fine. They're getting Cs or maybe occasional Bs." But it's relative to how smart you know your child is and how you think they should be doing.'

Do lower-than-expected school results matter? Haven't these children already effectively won the intellectual lottery? Obviously, their emotional welfare must be considered, along with that of every other child. But their underperformance is also Australia's loss, according to the Victorian parliamentary inquiry into gifted and talented children, which estimated between 10 and 50 per cent of gifted children were failing to reach their potential. 'Gifted students are our prospective leaders and innovators,' the inquiry's report said. 'In nurturing their talents, we are not only meeting their rights to access an appropriate education, but also ensuring that the future of our society is in good hands.'

An Australian Senate inquiry concluded in 1988 that gifted children were among the lowest priorities of all educationally disadvantaged groups. Jae Jung puts it this way: 'There's always this focus on the kids who are struggling and, of course, they need support. But just as they need support, the kids at the other end need support as well. The fact that they're not being appropriately supported is demonstrated by the long declining trend in the performance of Australian students at the top end in international assessments.' Australia has been slipping down the rankings of the Programme

A Senate inquiry concluded that gifted children were among the lowest priorities of all educationally disadvantaged groups.

for International Student Assessment tests, which are conducted once every three years, since its inception in 2000.

Isabella's first school did not believe that 'giftedness was a thing', says her mum. 'Therefore, every child was to finish the year at the same level. At first, Isabella desperately wanted to go to school. But she learned very quickly that it was easier to hide.' She was so bored and frustrated in class she ground down her front teeth 'They're actually damaged,' says Isabella.

After years of struggles, including a period of homeschooling, her parents finally enrolled Isabella in a private school's remote learning program where the class levels were matched to ability rather than age. Isabella is clearly much happier these days.

'It's a pretty good school,' she says. 'There're about six kids in my class and most of them are like me. It's very focused on individuals. It's pretty advanced stuff, too. It's not like normal schooling. So we get a lot done.'

Her advice to other gifted children? 'They're obviously going to be different and you have to acknowledge that. Also, it's not, like, a really good thing. So you shouldn't just think, *Oh, I'm so smart, like, everything's going to be easy.* But it's also not really a bad thing. I guess you get used to living with it.'

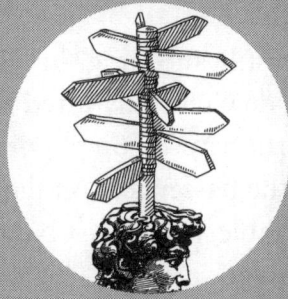

26

HOW DO YOU MAKE THE RIGHT DECISIONS IN LIFE?

When his plane's engine's failed, a pilot had to think fast. Here's how the rest of us can make tough calls, whether on marriage, moving house or matters of life and death.

Jackson Graham

On a clear, still morning in 2010, pilot Richard Champion de Crespigny lifted the nose of the A380 from the runway at Singapore's Changi Airport. Bound for Sydney, the passengers on flight QF32 barely had time to get comfortable when two bangs sounded and the plane began to shudder.

A turbine disc in one of the engines had exploded, breaking into shrapnel that ripped hundreds of holes in the Qantas aircraft. Shock ran through de Crespigny. Then the captain's decades of training kicked in. 'It could have easily gone the wrong way and the aircraft crashed,' he says. Simply following procedure would get the five pilots in the cockpit only so far—it might even put the 469 people on board in greater danger. The crew had to think independently of the aircraft's automated prompts and follow de Crespigny's lead. 'There are so many ways to make decisions, and on QF32 we used all of them,' he says.

Making decisions is somewhere between art and science. Flying a plane that experiences engine failure might be one extreme, but we all make life-changing calls at times. Some of them carry more weight than others: What career will I pursue? Should I stay or leave a relationship? Where should I live? Have kids or not? Is it time to retire? When we get stuck, our emotions and values, our plans and their pros and cons roll around in our heads like metal balls in a handheld maze.

Making decisions is somewhere between art and science.

Is there a good way to make big decisions? When should we rely on intuition? And what if we make a choice that we regret?

HOW DO YOU DEAL WITH MATTERS OF LIFE AND DEATH?

The cockpit shook, alarms blared and warning lights flashed red. Panic would overwhelm most of us but de Crespigny didn't hesitate. He switched off the autopilot and stopped the aircraft's climb.

Pilots are taught that the first 30 seconds of a crisis 'can be the difference between life and death', de Crespigny says. 'All I did was make sure the aircraft was flying and that we were safe.' In those critical first few seconds, you need to 'create time', he says. 'Don't get drawn into making any rash decisions.' Panic is a subconscious response that pilots train to overcome. 'Practising being out of your comfort zone is really good for protecting you from the fear response,' de Crespigny notes.

Although the plane was still in the air, the damage was severe; one engine had been destroyed, another three were malfunctioning. All but one of the 22 flight systems had problems, including the brakes and landing gear. The pilots put the plane in a holding pattern while they prepared for an emergency landing at Changi Airport.

One theory of how we think, popularised by Nobel Prize-winning psychologist Daniel Kahneman in his 2011 book *Thinking, Fast and Slow*, breaks our minds into two systems. The first uses mental shortcuts from prior experience. The other strains over slow, deliberate decisions. De Crespigny needed both to survive this crisis. 'When you're flying an aircraft, the 35 years of flying experience that I had gave me all these basic skills to operate subconsciously,' he says. 'It makes decision-making fairly quick—or you've at least thought through the processes.'

Still, he and his team thought carefully before following many of the aircraft's prompts; they ignored a direction to transfer fuel from one wing to another, for example, given there were holes in the fuel tank. 'If we had followed those checklists, we probably wouldn't have gotten back.'

Nearly two hours after the explosion, a 'magic silence' came over the flight deck as de Crespigny dropped the plane towards the runway, he recalls in his book *QF32*. It was still heavy with fuel for the journey and coming in too fast. A stall warning sounded metres from the ground. The A380 used nearly all four kilometres of tarmac to come to a halt. 'After we'd stopped, I had a moment of pleasure thinking the crisis was over but it wasn't,' he tells us. 'It was like a bushfire when the wind changes direction.'

In fact, the decision-making was entering another phase. Fuel was pouring out of one wing near the plane's hot brakes. One of the engines would not shut off. A complex question arose: should everyone on board be evacuated via emergency slides that would, among other risks, potentially injure the elderly and put passengers in close proximity to jet fuel that might ignite as they walked across the tarmac? 'We decided to keep the passengers on board,' de Crespigny tells us. He believed it was safer to wait for stairs to be brought to the plane. 'Passengers could have died if forced to evacuate,' he says. 'I would say 60 per cent of pilots would say we made the wrong decision not to evacuate.' But he kept returning to his training: 'Keep calm, don't rush, create time, never presume or assume and face only the danger in front.'

'Keep calm, don't rush, create time, never presume or assume and face only the danger in front.'

ARE THERE TECHNIQUES FOR MAKING BIG LIFE DECISIONS?

It was 'the day of days!', Charles Darwin wrote in his journal on 11 November 1838. His excitement wasn't due to a breakthrough in his theories of evolution. Rather, he'd asked his cousin, Emma Wedgwood, to marry him and she'd said yes. It would be a union of four decades, the rest of Darwin's life. Months earlier, though, he had been weighing marriage against his scientific career.

The most common important decisions people make involve their relationships, family, career, education, finances and home, says University of Technology Sydney associate professor Adrian Camilleri. He surveyed 658 people about their biggest decisions, finding these moments were often rare, uncertain, required contemplation, involved personal morals or values, affected multiple people and had long-term consequences. 'The more of those elements that are part of a decision, the bigger it is,' he says.

Darwin used a rudimentary tool to ponder marriage. He drew a line down the centre of a page and scribbled the pros and cons on either side. The upsides were children, companionship, love and play—all 'good for one's health'. But there were downsides, such as less 'conversation of clever men at clubs', the 'expense and anxiety' of children and 'perhaps quarrelling'. Darwin jotted his conclusion at the bottom of the page: 'Marry'.

Several experts told us that a 'pros and cons' list can be useful if the alternative is not thinking through a decision at all. In some ways, the pros and cons approach remains state of the art 'because the art hasn't advanced much', says Steven Johnson, the author of *Farsighted: How We Make the Decisions That Matter the Most*. 'This got more and more scandalous to me as I researched and wrote the book; no one had ever told me how to make a decision.'

However, a pros and cons list can reduce big decisions into either/or choices too early. Relationship concerns framed as 'stay or go' are one example, says Dr Rowan Burckhardt, director of the Sydney Couples Counselling Centre. In his experience, people who are no longer in love rarely find therapy beneficial but for others working through conflicts or differences, it can open up options. 'You think that you're facing two roads in front of you: one is separation and one is just putting up with this forever,' Burckhardt explains. 'But, in fact, there is a third path—you just don't even see it.'

> A pros and cons list can reduce big decisions into either/or choices too early.

Johnson adds another approach: considering all the possibilities. 'If you don't predict well, then you can't decide well.' Some businesses call it a 'pre-mortem'— foreseeing nightmare scenarios in advance of a decision. But Johnson thinks of it more as a form of storytelling that can plot positive or surprising outcomes too. 'The other important thing is having a diversity of storytellers' who can offer their perspectives, he says. 'People just make better, more far-sighted decisions when they make them in heterogeneous groups.'

Before making a critical decision about a checklist on QF32, de Crespigny consulted each member of his team, from the lowest-ranked pilot up, to make sure he hadn't overlooked anything. 'If you start at the bottom, everyone is going to have their say,' he explains. 'You don't get any groupthink and you certainly don't get peer pressure from someone being reticent to speak up.' He suggests using this technique at home, asking the youngest child's opinion first.

When facing a big decision, gather as much information as possible, says Ben Newell, a professor of behavioural science at UNSW. It sounds like a trite thing to say, but it's true, he tells us: if you can get as much evidence as possible in front of you, that will help you distinguish between options.

Next, Newell suggests, prioritise the most important aspect of the choice. When buying a house, for example, 'Is your budget the thing that's driving you, is it being in an area that's not going to be prone to floods?'

With the options and scenarios on the table, Johnson says, you can decide what you value most. '[Perhaps] having kids is much more important than conversations with clever men in clubs,' he says of Darwin's choice. 'They're both meaningful but one is way more important.' De Crespigny also keeps his personal values in mind. 'When you identify your values, it makes it easier for you to make a decision.'

That said, de Crespigny has struggled with plenty of decisions. For years he wanted to do the Cresta Run in Switzerland, a hair-raising kilometre-long plunge down an icy track on a skeleton toboggan, but he decided not to because of the risks (pilots are subject to regular physical tests). When his piloting career ended unexpectedly because of the pandemic, he decided that it was time. 'You go head-first at 80 miles an hour with your head two inches from the ice,' he says. 'Your risk appetite changes during life.'

When the time to make a decision comes, no single technique can tell you exactly which way to go. But the experts we spoke with said people were generally happier with their decision when they took a methodical approach and

People were generally happier with their decision when they took a methodical approach.

335

gave themselves enough time. 'Having some kind of structure is the most important thing,' Johnson says.

CAN 'GOING WITH YOUR GUT' OR 'SLEEPING ON IT' HELP?

We've all seen someone who doesn't need to mull over a decision, perhaps believing they 'just knew' what to do. They might have raised their hand at an auction for a house because the place had a clawfoot bathtub like their grandmother's, or they might have married the person of their dreams after just a few dates. How much should we let our subconscious do the deciding?

UNSW professor Joel Pearson, author of *The Intuition Toolkit*, says intuition—that subconscious response people feel in their body and can act on—comes from experience; using it in situations we rarely encounter is risky. 'If you become an expert in something in the workplace, that doesn't mean your intuition will transfer,' Pearson says. And don't rely on your gut if you're stressed, anxious or angry. 'You confuse what caused that emotion with the emotions for a decision. Wait until you've calmed down and you're in a clearer state.' Situations in which time is short 'and all the information is uncertain, ambiguous or limited are probably the most obvious cases to use intuition'.

After landing the A380, de Crespigny gave his phone number to every passenger, telling them to call if they had any difficulties with the aftermath of the ordeal. The decision to do so didn't take much thought. 'I truly wanted to make sure that they were happy when they were home and they weren't going to be traumatised,' he says. 'By accessing my emotional brain and coming up with *why* I have to do something, I'm not in conflict at all when I have to make decisions.'

Do sleep, dreaming and the unconscious help us make decisions? In de Crespigny's experience, dreaming helps him feel less inhibited in his thinking. 'When you dream you can be creative and think up solutions you never thought of when you're awake,' he says. 'It's really good to take a problem to bed.'

When it comes to making a tough decision, though, science hasn't proven that delegating the task to your subconscious has any benefits, says Newell. He has studied whether people make better decisions when distracted—say, while solving a word puzzle—compared to people who are focused only on the task at hand and found the latter group 'made just as good, if not better decisions'.

Still, there is something to be said for 'sleeping on it' or 'going for a walk' to let the mind rest rather than wander. 'Decision fatigue is a real thing,' notes Rachel Searston, a psychology researcher at the University of Adelaide. When he was US president, Barack Obama famously wore only grey or blue suits to cut down on trivial decisions. 'I don't want to make decisions about what I'm eating or wearing,' he told *Vanity Fair* in 2012. 'Because I have too many other decisions to make. You need to focus your decision-making energy. You need to routinise yourself.'

Searston has observed crime-scene fingerprint examiners who say they make better final judgements after coming back with fresh eyes the next day. 'We can do that in our everyday life too,' she says. Newell describes it like this: 'I stop and then I come back, but when I come back, I'm then engaging in that deliberative thinking again. I'm thinking about it again but I'm thinking about it in a different way and that then leads to a better solution or a better answer.'

WHAT IF WE REGRET OUR DECISIONS?

In the 51 years that Judith Viorst lived in her stately home in Washington, DC, she made a habit of walking on the wraparound porch and putting her arms around one of the Ionic columns. 'I love this house,' Judith, author of the 1986 bestseller *Necessary Losses*, would tell her husband Milton.

When they reached their nineties, Milton, a former Middle East correspondent for *The New Yorker*, struggled to climb the house's three flights of stairs and its Victorian-era foundations were showing their age. Then, when the couple were on their way to a medical appointment, Milton lost his balance and fell on top of Judith, breaking her pelvis. 'We started realising, we're old. No kidding,' she says.

Judith has given her fair share of thought to loss. 'We lose not only through death, but also by leaving and being left, by changing and letting go and moving on,' she wrote in *Necessary Losses*, a non-fiction book that explores the separations and losses that punctuate our lives. But many years later, downsizing to a retirement village was a loss that Judith hadn't foreseen for herself, she tells us. 'It was really like the five stages of grief.'

Her house was a time capsule of raising three children, hosting celebrations in a dining room that sat a couple of dozen people, and she and Milton working in their respective studies. (Milton, who died in 2022, wrote 10 books; Judith has written 43 so far.) To stay or go? 'It was a monumental decision,' she says.

At life's big forks in the road, Judith will first acknowledge the losses involved. 'You don't kid yourself or try not to think about it,' she explains. 'Say, "This is something I love, that I'm giving up or that's being taken away from me".

You sort of have to face the reality of the loss head on and then you can start to think about stages of life, passages of time.'

The experience of regret, says Adrian Camilleri, often strikes people when they haven't made a decision or it has been taken out of their hands. 'It's maybe not that they made the wrong decision or they would have made a different decision but it's the fact that they didn't go through that process of actively deciding.'

Regret often strikes people when they haven't made a decision or it has been taken out of their hands.

For Judith, leaving her house seemed impossible: it was full of furniture, 15,000 books and five decades of belongings. What to do with them all? 'All of these were decisions that we had to make,' she says. 'I don't think there was anything quite like this since when we decided to get married.'

With help from family and friends, she shortlisted four retirement villages, then chose one. 'It's really pretty, filled with sunlight,' Judith tells us. 'There were beautiful leaves fluttering in the breeze against all of our windows. What you're looking for is something that will make it easier to have made that decision.' When buyers began to inspect her house, she 'hated' them. 'Somebody said, "I wonder if we can get rid of those columns?".'

The right buyer eventually came along: a family similar to the Viorsts when they moved in all those years ago. Before she left the keys behind, Judith invited friends from her street to a 'farewell ceremony' where people told stories and reminisced about the house, and she's been back as a guest, touring a renovation that has repaired or replaced all the details she loved. 'It is absolutely beautiful; it's become their house, it's not my house anymore.'

Life in a two-bedroom apartment is certainly different, says Judith, and the retirement village isn't like the old neighbourhood. 'Everybody is old, you're not going to see a five-year-old learning to ride a bike.' But she has close friends and plenty of time to write. 'It is as satisfying as it could be.'

What she sees as peoples' biggest decisions will sound familiar to many of us: who to marry, what career to pursue, whether to have children and what personal values to uphold. There's no point in making a drama of it, though. 'That's called growing up,' Judith says. 'I think there are losses involved in making choices. But I don't think you go through life brooding over whether you went left instead of right. I think you make the best of it.'

About the journalists

Matthew Absalom-Wong is the national creative director at *The Age* and *The Sydney Morning Herald*. He has won various awards for his multimedia design, including two team Walkleys, and for his artwork, including two Melbourne Press Club Quill Awards.

Kate Aubusson is the health editor at *The Sydney Morning Herald* and an award-winning medical journalist.

Eryk Bagshaw is an investigative reporter for *The Sydney Morning Herald* and *The Age*. He was North Asia correspondent from 2021 to 2024 and won the Kennedy Award for Outstanding Foreign Correspondent in 2023.

Jamie Brown is a graphic artist at *The Age*. His infographics and illustrations have won awards at the Melbourne Press Club Quills and the international Society for News Design. He was a Walkley finalist for best artwork in 2018.

Daniel Ceng is an award-winning photojournalist in Asia. He has covered the war in Ukraine, tensions between China and Taiwan, and social and environmental issues in the Philippines and India.

Angus Dalton is a science reporter at *The Sydney Morning Herald* whose work appears in three editions of *The Best Australian Science Writing*. He won the inaugural *Herald* tennis tournament in 2023 and is campaigning for a pickleball spin-off.

Rachael Dexter is a city reporter at *The Age* covering suburban life and local council politics. She won Melbourne Press Club Quill Awards in 2020 and 2021 and produced the Walkley-nominated multimedia series *Invisible Crime: Australia's Sexual Assault Crisis* in 2019.

Billie Eder is a sports reporter covering rugby league at *The Sydney Morning Herald*. She holds a degree in political and social science. An animal lover, she previously worked as a rider and groom at a show-jumping stable on the NSW Central Coast.

Osman Faruqi was the national culture editor at *The Age* and *The Sydney Morning Herald*. He previously worked as a reporter and editor at the ABC where he won a Kennedy Award for audio reporting in 2019 and an Australian Podcast Award in 2021.

Sherryn Groch investigates crime for *The Age*. She was the inaugural explainer reporter for *The Age* and *The Sydney Morning Herald*, from 2020 to 2023. She won a Melbourne Press Club Quill Award in 2024 and a Young Walkley Award in 2020 while at *The Canberra Times*.

Patrick Hatch is a transport reporter at *The Age* who previously reported for the national business desk, including on the aviation industry.

Carla Jaeger investigates crime and corruption for *The Age*. She previously worked on the *Age* sports desk covering the power, politics and business of sport, for which she was named Young Journalist of the Year at the 2023 Melbourne Press Club Quill Awards.

Samantha Selinger-Morris hosts *The Morning Edition*, the daily news podcast from *The Sydney Morning Herald* and *The Age*. She worked for *The Globe and Mail* in Toronto before moving to Sydney 26 years ago, where she wrote for the ABC and SBS before joining *The Sydney Morning Herald* and *The Age* as a feature writer.

Jewel Topsfield was the social affairs editor for *The Age*, with previous roles including education editor and Melbourne editor. Indonesia correspondent from 2015 to 2018, she won a Walkley Award for international journalism in 2015 and the Lowy Institute Media Award in 2016.

Matt Wade is a senior writer who covers economics, politics and demography. He has twice won the Australian Council for International Development Media Award for coverage on Africa and the Pacific. A correspondent in India for *The Sydney Morning Herald* and *The Age* between 2007 and 2011, he was previously the *Herald*'s economics correspondent in Canberra.

Damien Woolnough is the national style editor for *The Age* and *The Sydney Morning Herald*. He has worked for *Vogue Australia*, was the launch editor of vogue.com.au and was the fashion editor for *The Australian*. Damien has written for many fashion magazines, from *Harper's Bazaar* to *The Australian Women's Weekly*.

Thank you . . .
To wrap your head around a complex subject, draw out its tensions and intrigues, then explain them accurately and eloquently—in just days—is not a task for the faint-hearted. Explainer reporters Jackson Graham and Angus Holland bring great rigour, integrity and good humour to their jobs; as did their predecessor Sherryn Groch, who blazed a trail for engrossing explanations as the first reporter on our desk. Other journalists from our newsrooms took time from their busy schedules to write pieces in this anthology; thank you to them and their editors, who made this possible, as well as to all the experts quoted in this book, many of whom graciously fielded several rounds of questions as we checked and double-checked our facts. Thanks also to brilliant interns Thomas Bailey and Annie Holland. My boss, Chris Paine, has been supportive all the way; and so have the sub-editors, graphic artists, librarians, videographers and other editors who help us to deliver high-quality explainers every week. Finally, thank you to the publisher of this book, Sally Heath, and her team at Allen & Unwin, especially senior editor Samantha Kent, for enabling us to present our explainer journalism in such fine form in this book.